Mosaic Life
Moving and Provocative
Autobiographic Anthology: 1

Created & Written
by
"Frank Julius" Csenki

Published by Frank Julius Books

www.frankjulius.com

Frank Julius Books

© Copyright January 2016
All rights reserved, reproduction in any format is prohibited

ISBN: 978-0-9950178-5-6

This is a collection of short stories of true events, told as close to reality as memory serves. In some cases, the names and characters represented have been changed. Any resemblances to persons living or dead are entirely coincidental unless otherwise stated.

MOSAIC LIFE TILES
MOVING AND PROVOCATIVE

AUTOBIOGRAPHIC ANTHOLOGY : 1

The Occurrences of Tiles

Okay, so this is the starting point. As you know, the title of this book is; "Mosaic Life Tiles."

These tiles to be clear, are individual experiences and elements that touch upon the sides of other tiles that then come together in the composition of each story.

This is a book of short stories. You will find that one story will have a tendency to link to another story, by association with tiles past and tiles future depending on the placement within the mosaic composition that makes up my fragmented life.

It is with luck, guidance, and advice (from family and close friends) and careful thinking on my part sometimes, *"he said, with a grin,"* that allowed this fragmentation to remain glued together, securing things in place and holding my life together.

No doubt as time marches on, the glue will dry and no longer hold fast all the tiles. I imagine that will become self-evident as my health deteriorates, my bones become brittle; my muscles will start to strain and ache, as atrophy sets in and the wonder of old age creeps up and takes ahold of me.

I have no doubt that I will awaken one morning and look in the mirror as I do every day, then say, "holy crap whatever happened to me!" And I will be seventy-eight years old seemingly all of a sudden; overnight!

Is that how it happens? I have often tried figuring out the passage of time, and how it marches on, but always seems to stay in the present.

Yesterday is dead and gone, and tomorrow never comes; as the song says. I believe that was a Kris Kristofferson tune, you know the one I'm referring to, and so yes, that would then logically only ever leave the present to deal with.

The present to me has always been the past, now and the future all rolled up into one, and thus I will be looking in that mirror on that morning when I am seventy-eight and ask, "how the hell did I get here?"

Hopefully the crumbling of my "life tiles" will be a slow process, but for now, the tiles hold tight, and the mosaic is clearly seen. But as my memory will fade, so will the tiles become more obscure and opaque, the cracks between them widen, and the tiles will chip away from the passage of time.

But like I say, for now, they hold tight and perhaps I can do them all justice in bringing them to life in these words, and allow them forever to hold tight in print.

Here I offer you a close-up to each one in detailed drilled down insight and fashion.

Now having said that it does not mean that I'm about to start at the beginning of this life of mine, but rather tell you about my life's interesting happenings; my life's "Mosaic Life Tiles."

This would then cover the period from now; being defined as the time of your choosing to read this and maybe all the way back to the roots of my memories, which I presently put forth here. See how that *presently* inserted itself into my past again?

The funny thing about writing is that although there is a fundamental platform governing the setting, the characters, and the plot, writing does have a way of taking on its own life, directing the author into areas sometimes hard to avoid, and needing exploration. The trick is to bring things back on track and remain focused, driving to the conclusion, and a satisfactory conclusion at that.

I'll be talking to you about things that stand out in a man's life, a teenager's life, a little boy's life. With some luck, you will be able to relate to similar experiences of your own, and if not similar, then my aim will be to enrich your emotions with heightened levels of imagination from my mind's eye to yours.

I firmly believe that the theatre of the mind is the most vivid, absorbing and spectacular venue of the human experience as evidenced by your own incredible dreams.

For those of you of the opposite sex, who happen to read the words contained within these pages, I trust that you will with some caution and appreciation open your mind to the workings of a male mind and accept both the pros and cons of my delivery. So without further ado, I will tell you about the experiences that created my mosaic life tiles.

These experiences I will convey in the form of short stories, an anthology on the first series of life tiles. I will be sharing with you the unusual, the incredible, the unbelievable and the thought-provoking realities that I am hoping will trigger amazing connections to experiences of your own, and with that, my life tiles can momentarily interchange places from within my mosaic makeup to yours.

We've all had incredible life experiences that are unique and cannot be matched because only you can appreciate the way in which it happened to you, in a very personal way.

But sharing those experiences adds to the collective human bonding of living our lives, learning from each other, reveling in joy, finding and offering empathy in despair, thus making the connection that unites us for a momentary interchange of life tiles. Some of those tiles to be found in this first anthology, and I give you:

Mosaic Life Tiles
Moving and Provocative.

***

Me: 1 – Company: 0

My last day in Moscow had arrived on April 6^{th,} 1993. I had such mixed feelings about everything. On the one hand I was leaving behind one of the most incredible life experiences I had up to this point of my life.

I had just completed seven months living in the capital of Russia, and let me tell you, from the bottom of my heart; it was with great sadness that I was facing this morning and the day to come. And yet, I was excited and relieved to be leaving Russia and returning to Canada.

How does one balance such opposite feelings of the heart and mind? Yes, this is about my job interview to Vienna, but it starts here in Moscow.

Just to qualify things up front, I would not be going to my upcoming job interview in Vienna if I hadn't been in Moscow in the first place. It all came about because of my time in Russia.

My last day in Moscow came about after having completed two separate tours so to speak in military terms, but mine, of course, was civilian.

My consulting assignment for the Moscow Aerostar Hotel had come to a close and this day I would be returning to Canada and subsequently to the USA where I lived; in Florida.

Having spent the last seven months in Russia opened my eyes in many ways. How do I explain? Let me try like this...

The city, its people, its atmosphere, its way of life, its opulence, its gutter poverty, its emergence from communism, all this; that the new Russian frontier was, had become very amplified parts of my life.

My understanding of daily existence, all of these things, I lived daily through the eyes and love of my Russian girlfriend; Irina.

I was leaving Russia, I was leaving her, I loved her, and that morning had started with tears in my eyes. This life tile will be revisited another time, but I wanted to set the picture.

After having spent the time I did in Russia and working in that chaotic business environment, I happen to be one of those individuals who had an understanding of what was going on, what to expect, with a healthy dose of caution added to all agreements made.

During my seven months of exposure to the new Russian frontier, it allowed me to both witness and participate in this budding emergence of capitalism and how to go about in managing and getting along with something totally familiar to me, but totally confusing to the average Muscovite.

This truly were unique circumstances; not a lot of people were exposed to, not in the way that I was.

It was just after the fall of communism in Russia and free enterprise had been born overnight.

After I had left Russia that day, my seven months of experience and know-how gained had launched me into a select group of individuals who were sought out by western companies, which I was to find out about a few weeks later.

These western companies were looking to establish partnerships in Russia and the CIS *(Commonwealth of independent states being the newly formed countries of Belarus, Georgia, and many more with the breakup of Soviet Russia)* and get the economy flowing.

So I left Moscow and flew to Canada. My mother and brother and his family lived in Ontario. My home was in Port St. Lucie Florida, but I flew back to Canada because I was on my way to Alaska in four days time to do another assignment, a short-term two-month assignment in Dutch Harbor in the Aleutian Islands, to open a new hotel. I did that.

◇◇◇

    Just after completing that Alaska assignment I found myself talking to an Austrian Hotel company.
    This company tracked me down to interview me for a job in one of their hotels located in Tiblisi Georgia. Formerly part of the Soviet Union, now one of the nations making up the CIS as a result of the breakup of the Soviet Union.
    So now with the scene and situation set, I finished my assignment in Alaska (another life tile to be visited) and flew back to my mother's place in Hamilton Ontario to visit for a few days before I finally and eventually was to drive back to Florida and go home!
    All this time long, I had my car parked at my mother's place in Hamilton Ontario because I flew to Moscow originally from Toronto, and left my car in Canada after having driven up from Florida.
    Ok, so the story goes on. I finish my gig in Alaska, find myself visiting my mother in Hamilton and the phone rings.

When I answered the phone, much to my surprise, it was the corporate controller for Marco Polo Hotels calling me from their head offices located in Vienna Austria.

I am not sure how he obtained my mother's phone number, but he did! I am taking a stab in the dark, and thinking he had called the hotel in Moscow where I had worked a few months back and they gave him my Canada contact number. That must be it. So, here's how it went.

Hello, I answered the phone.

"Yes good day, I am calling from Vienna Austria for Mr. Julius." The male voice on the other end said.

Well, I was pretty surprised, but right away I knew it had to be something to do with a job in Europe, he was calling from Austria after all.

"Yes, this is Frank Julius."

"Mr. Julius, my name is Reinhold Schmidt, I am the corporate controller for Marco Polo Hotels, our head office is here in Vienna. I will get to the point immediately." He said in an Austrian accent sounding like Arnold Schwartzenhager.

I felt like saying and almost replied."Well thanks for the call Arni," but of course I didn't, but if you had heard

Reinhold, you would be very tempted as I was.

I have this crazy sense of humor sometimes that forces me to say something totally inappropriate in the most serious of times, just to take the edge off but I held back.

Anyway, he continued.

"The reason for this call Mr. Julius is to invite you to Vienna since we at Marco Polo Hotels are interested in talking to you about an opportunity with one of our properties."

I was right.

I had this feeling immediately upon answering the phone. I have to tell you, I have this sixth sense thing going on when it comes to these things, specifically calls just out of the blue, which in most cases were and are job opportunity related, and usually concerning job interviews.

So I responded.

"Thank you, Mr. Schmidt, for your call. I am familiar with your company and know of your hotels.

I have just completed an assignment only this week, and your call is most interesting to me since I was about to search for new opportunities in the coming weeks."

Schmidt then continued on.

"Mr. Julius, we are prepared to fly you to Vienna, provide you with accommodation for the time you are here and cover your expenses, how does that sound?" Schmidt asked.

Then it was my turn.

"First, thank you for you call and please call me Frank. I would be very happy to move onto the next step, but of course, we need to discuss the position and to save time, let's get to the point right away and discuss the position, location, and salary; the entire package, does that sound fair Mr. Schmidt?"

"Yes Frank, that is fine. The position would be hotel controller. I had become aware of your background and your contribution to the success of the Moscow Aerostar Hotel during your time with them. You may know that we operate a hotel in Moscow and one of our executives there who came to know you, spoke highly of you."

At this point, I had to cut in.

"Thank you for the compliment Reinhold."

I figured he already addressed me by my first name; now it was my turn to return the courtesy.

"So, tell me Reinhold, which property is this position for?"

He responded.

"It's for our hotel ski resort just outside of Tiblisi Georgia."

My mind at that point switched into high gear, and I ran the scenario through in a matter of seconds.

The first being that I already knew of this company and had heard that they were a little on the "cheap side" when it came to salary and they skimped wherever they could.

(In the interest of full disclosure; as of the writing of this book, this which was located in Vienna is now defunct, and the new Marco Polo Hotel company located in Hong Kong has nothing to do with the Marco Polo Hotels I write and refer to in this book.)

I heard they skimped on certain benefits to their executives. But at this point I was still okay with it; it was just a phone call for now, but I kept the thought under my hat.

The second issue was, in fact, the location that being Tiblisi Georgia. Truth be told, I had no desire in going to Tiblisi Georgia. I recalled hearing that Eduard Shevardnadze who had been Gorbachev's right-hand man and his Foreign Affairs Minister during the decline of the Soviet Union was running for The Presidency of Georgia and stability of the nation was not assured. Georgia had become an independent nation with the recent collapse of the Soviet

Union, and it was having a tough time in establishing stability as a nation transitioning from socialism to free enterprise and capitalism.

There was much turmoil and corruption in the country.

I had just finished experiencing growing pains of the Russian economy and wasn't looking forward to dealing with corruption gone rampant in Georgia. On top of that, having lived in Russia for as long as I had, I learned that other expatriates who had taken jobs in Georgia were experiencing way too much difficulty in fitting in with Georgian society.

But there was one aspect of this conversation with Reinhold that did peak my interest, and that was the mention of "Fischamend."

Reinhold said that I would be required to stay in the town of Fischamend just a half hour or so outside of Vienna. This part was great!

I will never as long as I live, ever understand how fate and circumstance seem to come about from time to time in one's life that demand a follow-up, for me "Fischamend" was the draw that I could not refuse. I guess that is what fate is all about, something uncontrollable and meant to be.

I will get to that in a bit.

I continued on.

"Reinhold, so that we are both on the same page, I would be happy to come over for the interview, but before I commit, I will need to know if your compensation package is in line with my needs. If you could elaborate on that, I will then be in a position to make an immediate decision."

At this point I knew I was going, come hell or high water, because I wanted to go to Fischamend. I only asked for this to be settled so that Reinhold had a certain comfort zone. But make no mistake; I didn't care what the compensation package was.

Reinhold had no problem disclosing this up front. Naturally, he didn't want to take the chance of having me fly over, only to find out later that his offer was much too low and my interview could have been offered to someone else instead more in line with his company's offer.

The thing about these sorts of opportunities was that the position and location had to be a good match for the candidate, and there weren't a whole lot of candidates out there.

One might well think that this would be a very coveted position and opportunity with a long line of candidates eager to experience such an adventure in career development, but that was not the case. Let me explain some of the difficulties in securing the right person.

First, you have to be available, not in six months or even two, but now.

Second, the person needs to be efficient, adaptable, and preferably single. Single is good for several reasons. First, the paperwork is less cumbersome; work permits, visa, passport and the list goes on.

Second, with a single person, there is significantly less risk regarding adaptation success.

But to find a single person, that would also mean someone without a significant other, and that is rather difficult. To hire someone who has the smarts, the ability, the experience and the managerial expertise, well, such an individual would normally already be in their mid-thirties. It does, in fact, take a few good years in the business to attain a position of executive level status. One does not rise to the capacity of financial controller right out of college.

There are a number of roads one must have traveled to arrive at such a position, especially in the hotel business which is so diverse. Running a resort hotel operation is similar to managing a small town. The resort could at any one time have over a thousand guests participating in a variety of activities, requiring great management attention and know how.

By the time an individual is qualified to run the finance department of such a facility, that person is usually in their mid-thirties, and that usually entails having a growing family. Wife, husband, kids, dogs, cats, and now you have a recipe for a huge tactical and strategic nightmare.

To move a family into an unknown environment is just much too complicated. So we are back to the *single* person. That person, is very rare, no attachments, willing to travel and is an expert in the area you are looking to fill.

I was one of those individuals. I already had over fifteen years of international experience in the hotel business having worked in Canada, the USA, Bermuda, Bahamas and other Caribbean islands as well as Russia.

The fact that I was in Russia just soon after the end of the Soviet Union with the birth of free enterprise during the "transition" phase, well that was something very unique. Yes, I happened to be at that time of my life totally single, without a significant other… well, sort of.

You see, I did leave a love behind in Russia, her name was Irina with whom I was madly in love, but I had to leave, and I had no choice but to leave her behind.

As events would have it, and as the situation and circumstances allowed, we were, in fact, able to get back together several times with her flying to the USA and Canada many times afterward, but that I will touch upon later.

Back to my phone call with Reinhold, he explained the salary and overall compensation package. To be honest, it was just barely borderline for what I would normally accept. It was to be sure on the low end of the scale, but then again, there may be other aspects configuring into the equation.

Those variables being such as; the type and quality of accommodation the company was willing to throw in, and the perks in-house, at the resort, would go a long way as well, but those things were still to be identified and discussed. However, for now, I was good. So I accepted Reinhold's offer over the phone.

"Reinhold, I believe that sounds all fine to me at this point." I said and continued, "I will need to have a written offer before I commit to flying over. I will need you to fax me your invitation on your company letterhead, stating your agreement to full reimbursement for all incurred travel expenses; from my point of origin here in Canada, to your corporate offices and return to my point of origin in Canada.

This reimbursement to be paid in full, in cash, U.S. Dollars before I leave your offices returning to Canada. Is that something you can do today Reinhold?" I asked.

Reinhold said, "yes Frank, I can do that. Please fax back your arrival date and time, so we can arrange to make reservations for you in Fischamend at the Sleep Inn. That's one of our properties; you will be staying there."

So that was it, I would be flying to Vienna in the next couple of days, just as soon as I could make arrangements.

What was the real reason that I was going? The town of Fischamend was the reason.

You see, when I was just five years old; little guy that I was, my mother, father and I; (my brother Jim, hadn't been born yet, he was still living inside of my Mom), we ran for our lives. Yes, we ran through the woods, bushes, forest, farms, and mine fields as escaping refugees from Hungary.

That was in 1956, running for our lives, escaping communism and the terror that filled our homeland.

The Hungarian revolution was under way just then in November of 1956. That incident had Russian tanks rolling into Budapest to put down the uprising of the Hungarian people; who had had more than enough of their Russian communist overlords.

My father had been a freedom-fighter, and news from his fellow freedom fighters had reached him that he was a marked man.

He was wanted, and being hunted by the Russian Army and the AVO; Hungarian KGB, (pro-Russian Hungarian secret police) and that it was only a matter of hours before our house would be stormed, my father arrested and taken away. That would mean him being tortured and never to be heard from again.

We had only a few remaining hours to gather some meager belongings, meaning just coats and food for our escape.

We hurried and then set out to make our way to freedom, running for the border to Austria before my father was apprehended, my mother imprisoned, and who knows what fate would have awaited me.

The problem was however; that the border wasn't down the street, it was several days away and escaping, fleeing Hungarians were being apprehended everywhere by the AVO.

So we ran for it, and four days later, with the grace of God we made it safely across the Hungarian/Austrian border to the town of Fischamend, four days later.

This phone call from Reinhold awakened a point of history in my life that now beat the drums of return. I just had to go back and see this town. To be in it again, to smell once again, the first air of freedom that filled my five-year-old lungs back in 1956. The first day in my five years of life that I truly knew I need not be afraid anymore.

I knew that because my mother and father told me I didn't have to be afraid any longer. I could go outside and play as long as I wanted to. No more would I hear the loud thunder of cannons and mortar fire. No more would my Dad need to jump on top of me all of a sudden when hearing the deafening sound of incoming artillery.

I was only five, but I knew that something very important had changed for us. That change was to be found in Fischamend.

When Reinhold called, I was forty-two years of age, but after thirty-seven years, I still held vivid memories of this town Fischamend.

This town, where people greeted us with oranges, bananas, chocolate and chewing gum! Oranges and bananas, two fruits I had never tasted before in my life, and what was this wonderful delight, called "chewing gum?"

I can still recall the sensation in my mouth at the first taste of biting into this very strange but wonderful thing and then having the time of my life, just fun chewing it.

Ah, I was a five-year-old kid, simple things were huge for me. Oranges, bananas, and chewing gum…those had suddenly become heavenly items in my new world.

Upon arriving as escaping refugees to Austria, we had nothing. I mean absolutely nothing. We had eaten the bag of apples and the pair of dry Hungarian sausages my mother brought along, but that was it.

For four days we barely ate, we drank from streams. For four days and nights, I didn't say hardly a word.

My father and mother had told me to be very very quiet and never to let go of my father's or mother's hand in case we had to all of a sudden run.

I was just a little man, but I knew something very frightening was happening to us. I did as I was told. I remember my mother crying silent tears many a time during our escape. (This story will be another life tile to explore in depth at a later time).

Arriving in Fischamend, into freedom and the welcoming arms of the wonderful Austrian people, was like arriving in heaven.

We were free. The people, the town of Fischamend poured forth their generosity. We had warmth, we had food, we had safety and most importantly which I had yet to understand or realize, our family now had a future. An uncertain future to be sure, but a future in freedom, and that's what life is all about; freedom.

I didn't know it at the time of course, but many years later, after becoming a resident of the USA; no truer words ever rang more clearly in my mind, than those spoken by former Virginia Governor Patrick Henry in 1775, when he said:

"Give me liberty or give me death," upon addressing the need to take up arms in the revolutionary war against England.

Yes, as an adult now, I subscribe to those noble words. I am Canadian now, but some things American, I identify with closely.

So, back to my story. I am now on board Air Canada flying to Chicago from where I am to connect to; I believe it was Lauda Airlines.

Yup, the same airlines that had the tragic crash a few years later.

Arriving in Chicago, I get all messed up at the terminal for some reason or other; there was an issue with my luggage. With Chicago being my initial port of entry into the US, my luggage had to be unloaded and inspected by US Customs before it could go to my Lauda Air connecting flight and I end up almost missing my flight to Vienna.

Luckily I managed to scoop a decent seat close to the front of the plane, and I am good.

Arriving in Austria about nine hours later, I have this overwhelming feeling of nostalgia and history.

No, there wasn't anyone there at the airport to meet me. I took a cab from the airport to the central train station and then figured out how to select and find the train that would take me from central Vienna out into the countryside and to the town of Fischamend.

My interview was to be on a Monday. I made certain to arrive on Friday, Like that I had the whole weekend to myself to enjoy my Vienna visit.

It took me a while in figuring out just where I was in terms of my mind's eye zeroing in on Vienna's geographical location within Austria and its proximity to the Hungarian border.

Vienna straddles the Danube River (ah, the Blue Danube, always Johann Straus come to mind and the score for 2001 A Space Odyssey, but I diverge) anyway, just 58 kilometers southeast of Vienna is the Hungarian border.

Fischamend is roughly twenty kilometers outside of Vienna sitting close to the Danube. So once I figured that out, I more-less knew where my legs had me standing on planet earth.

Although we did make it across the border on our run for freedom, we, in fact, were transferred to Fischamend a day or two later after we initially made our presence known to a local farmer in Austria, who happened to be our first Austrian contact.

His farm sat right on the Austrian-Hungarian border. He had put our small starving family up for a couple of days, generously making us plenty of home cooked food and provided comforting shelter in his farmhouse until the Austrian authorities came to pick us up and deliver us to Fischamend.

There we were housed for several weeks and taken care of by the wonderful people in town and the Red Cross. (just a little more background there to bring things into line).

Then as things sometimes happen in life, fate intervened, and it turned out that the pastor of the local church was a distant relative of my mother's and my father also as it turns out had a distant relative in town.

This all went back perhaps a hundred plus years to the time when the Austro-Hungarian Empire existed and controlled a huge chunk of Europe from the German border to Russia and down to the Adriatic Sea.

The most incredible thing about it all with these newly discovered distant relations was that we had absolutely no idea that these wonderful people were related to us; actually lived in Fischamend until of course, we started talking to town's people.

They were curious as to who the new Hungarian refugees were in town, and once the grapevine of gossip and information started to flow, well it turned out that it proved to be for our benefit.

From that point on, my Dad was offered employment, we were offered an apartment for free, and a new life free Austria.

But no, my Mom and Dad were hell-bent on getting as far away from the Russian aggressors and communism as possible. Our aim was to find freedom and a new life in North America. It might too, could have been Brazil, but also is another "life tile," nevertheless; we ended up in Canada.

Finding the right train to board in Vienna that would take me to Fischamend turned out to be a piece of cake; it was so easy. Here I was, just having arrived, grabbed a cab to the train station and within ten minutes or so I knew how to get to where I wanted to go. Trains were frequent.

The signage was excellent. All one had to do was find the name of the town you wanted. Once you located the town on the Marquis; the train number was beside it.

You then matched the train number to the departing platform and departure times, and Bob's your uncle.

The train service was about once every couple of hours throughout the day. I had lucked out because I had only a few minutes to wait. It came, I boarded, and in a half hour, I was in my hotel room in Fischamend.

Upon arriving via train to Fischamend, my main aim was just to get to my hotel, freshen up and prepare to explore the town and most importantly bring hidden memories back to life.

It was still early in the day. It had been an overnight flight across the Atlantic. I managed to grab decent shuteye on the flight, and I was quite awake and energetic upon my arrival in Fischamend. So that is what I did. I checked into the small Sleep Inn, had a shower, shaved and prepared myself to visit my past, here in the present.

The past was waiting for me just outside my door in the coming few minutes of my future.

(Here we go again, with my past, present, and future dilemma, can't get away from it.)

My hotel; which was really just a motel, no restaurant or bar was about a mile or so from the center of Fischamend.

I grabbed my backpack and headed out from my hotel and down the roadside into town. As I walked for the first time after thirty-seven years on this soil of our family's first taste of freedom, I was overcome with emotion.

There was nothing I could do; my eyes would not hold back and tears formed, but those tears I recall were tears of joy, yet laced with memories of terror and fright.

I remembered my mother and father, just how desperate they both were in somehow making their way across the border, a border that had already been seeded with mines.

Each and every step we took was a huge leap of faith, hoping and praying we wouldn't be blown up to smithereens or even worse, wounded or maimed and having to be left behind or captured and then tortured.

Yes, these thoughts ran through my mind now at the age of 42. I confess, that those circumstances I did not fully understand being only five years old but make no mistake damn it I knew something was up, and it was not good.

But I knew with the beating of my heart as I walked down the road in Fischamend that those thoughts were very real and present with every breath and step that my mother and father took as we made our way across the minefield that separated the ground beneath us from totalitarianism and freedom.

So I walked, and I remember reaching into my backpack and taking out some tissue to wipe my face. The farther I walked into Fischamend, the better I started to feel. A certain peace and tranquility enveloped my walk.

The sun was shining, and as I walked alone down the road into town, people outside their houses; some folks waved to me. It felt almost as if they recognized me or something, I waved back, and I felt good. It had been a long time; thirty-seven years.

Some of my five-year-old memory I retained and could recall quite well. There were some things about Fischamend I still remembered.

One was the church in town, and the other was an incident that happened to me; a trick played on me by my father and my Godfather who also happened to meet up with us in Fischamend. He had also escaped Hungary with his wife.

Another was sledding in winter. I wasn't looking to relive any of those things; it was just comforting to think back on the time and to know that I was walking the streets now where those things happened.

The incident I had just mentioned, "the trick" my dad and my Godfather played on me. Well, I cannot remember of course which Gasthaus it was, but my Dad would have some beers with friends in town at the Gasthaus.

I would often ask my Dad where he was going, and one day his friend; my Godfather, told me that they were both going to the "switchblade club," and that it was a very dangerous place, full of rascals and hooligans.

Well, that sounded very very appealing to me. I was impressed that my dad and my Godfather would be going to such a place, and I wanted to tag along.

I was very gullible, and I believed everything that I was told and this to me was epic!

So they played along and told me that I could come along, but I would need to be very careful, and I needed protection from the rascals and hooligans.

My Godfather, (have no idea what he was thinking) gave me a little knife, a little folding knife, but it was a knife!

My dad for some odd reason thought this was funny I guess, and went along with it, kind of egging me on to see what I was made of, I guess.

Was I a tough little guy underneath or was I a sissy or weakling? Perhaps that might have been his reasoning, not sure.

Anyway, off we go to the switchblade club, in my mind.

We arrive, find a table and sit down. There were lots and lots of men around in the big beer hall; I remember all drinking beer and having a grand old time.

Well, little ole me; I was sort of on edge. I had no idea when one of the rascals or hooligans was going to come over and make trouble for us.

But I was convinced, that something bad was going to happen for sure, and boy-oh-boy was I ever ready!

Every man that walked by our table, I was watching like a hawk. I had my pocket knife in my hand; I was holding onto it already opened but in my pocket and I was ready in case anything was about to happen.

Well, wouldn't you know it? We were there no more than about three minutes. I remember the waiter had brought steins of beers over to my Dad's table.

I was sitting between my Dad and Godfather, watching all the goings on like a hawk, and on edge…like really on edge.

We had been in town I guess for a good few weeks by now, and my dad had gotten to know some of the town's people, especially his drinking buddies at the Gasthaus.

Well, as events would have it, there was this guy who spotted my Dad and was walking over towards our table.

I spotted him already before my father had. I was nervous; I didn't know this man and he was moving in on our table fast!

He extended his hand, reaching out over the table wanting to shake my dad's hand. I figured he was going for my dad's throat, and I suddenly jerked my hand out of my pocket and in the flash of an eye, I cut his finger!

Okay, so there was the switchblade club in action!

The poor guy snapped back his hand and didn't know what to do. My dad and my Godfather killed themselves laughing while the poor man was nursing his finger and totally confused. Oh, fuck it was funny!

I know now that it might sound like it was a very cruel thing to play on me, and for that matter, I could have stabbed the poor guy, but as things turned out, it was just a little cut, and the knife was taken away from me.

Good thing! Maybe it was one of those "had to be there" events, but thinking back on it now, I break up laughing. No, I was no sissy, I was ready to do battle!

So, on I walk into town. Some of it kind of looks familiar but most does not. The one thing that I do remember was this feeling of knowing I was in the right place; it just felt right.

Many years had passed. The town no doubt had changed, grew and had become modern. I walked around for quite a while searching for a connection, a landmark, something I could positively identify or something that would bring forth a certain memory, something I could touch.

I didn't find it, but I did find something of even greater value and significance. Something that will forever stay with me now, and it made my trip to Austria and Fischamend all the worthwhile. It was after midday, and I was looking for a restaurant. I found one that looked very nice from the outside. A Chalet type of building, and I walked in.

There were a few people inside not many; lunch hour was just finishing up. There were tables, and there was also a bar. I chose to sit at the bar.

The bartender came over, we greeted one another, I only spoke English of course, and I spoke Hungarian as well, still do, almost fluently.

I spoke up in English, and he also spoke English, so that was good. He right away knew I was a tourist in his mind, so after a bit, he came over, I ordered my food, and when it came, shortly after that, he came over to chit chat a bit.

I suppose he was curious where I was from, or perhaps he was just interested in talking. So I told him.

I told him that I was in town searching for my roots after we escaped Hungary and how Fischamend was so significant in our freedom.

The bartender then said. "Hold on a bit; I have someone you should meet."

He went out into the back, and two older people came out, a gentleman and a woman; husband, and wife, owners of the restaurant.

The bartender was their son. The owners came around to the front of the bar and walked up beside me. Much to my surprise, in Hungarian, the older gentleman said, "so you are one of our refugees my son tells us."

Oh my God, I was so surprised and overwhelmed. I just looked at these two wonderful and gracious people who were part of the great generosity that they showered upon me, my mom and my dad back in 1956.

They invited me to move to a regular table, where the three of us sat and talked for a while. It was so wonderful; it was magic.

We talked about the revolution in Hungary. We talked about the role that Austria played in welcoming us refugees, and we talked about how their backgrounds, the restaurant owners' were of Hungarian heritage, and they had always spoken both Hungarian and German at home, and now they spoke some English as well.

We talked a good while; I was so happy. In the end, we all hugged, and once again, for the life of me I had tears in my eyes. I left that restaurant but my day was made.

My trip to Austria was already a success, and I hadn't even been to my job interview yet. What a great adventure this was turning out to be!

My father had died almost three years before, but as I walked out of the restaurant, both my father and my mother were with me that day. I felt good; I had found something that day; found that I could still believe in people and that goodness was everywhere if you just looked for it.

It was only Friday afternoon; I had the whole weekend still to visit and breathe in Vienna.

The following morning I boarded the train in Fischamend and ventured into Vienna. I spent the next two days discovering this beautiful and amazing city, visiting the sites, the cathedrals, marveling at the architecture while eating pastries at sidewalk cafés and people watching.

It turned out to be a very relaxing and fulfilling weekend knowing that I was breathing the same air as did Beethoven, Strauss, Schubert, and Mozart. And I had lucked out because the sun shined both days and I became a tourist.

Monday morning was interview day with Reinhold. In my mind, I was already satisfied with my trip considering it to be a total success, with it being much more than I ever expected, but now it was my turn to meet my obligation and do what Reinhold had me come for to Austria.

I cannot recall the name of the street their head offices were located on, but I had a bitch of a time finding the address. I found the street quite easily, but the buildings on the street were more-less all joined which seemed to me like a block long of row housing connected to one another. But these buildings were all offices and businesses; and good luck in finding the right door.

Every few feet there were doors, and only some of the doors seemed to have numbers beside or on top. Now this was a major hotel company head office I was trying to find, and it seemed to me like I was looking for doors leading to mom and pop types of storefronts; just wasn't making any sense to me, but there it was.

I finally found the door with the corresponding address. Sure enough, I opened it, walked up a flight of stairs, and there was this long hallway with expansive corporate offices. It felt to me like I had just arrived in a different building than where the stairs lead me to, was very strange.

I introduce myself to the receptionist, and a few seconds later, Reinhold comes to meet me. Tall Austrian, no smile, very business like, we shake hands, and he asks me to follow him.

We walk into a glass enclosure, a board room that is glassed in and everything is visible from outside and inside, only the glass walls separate things.

So there is a boardroom table, with probably twelve chairs or so around it. Reinhold extends the invitation for me to have a seat, and I recall he asks if I would like a cup of coffee. I accept, and he arranges for the coffee to be brought in and at the same time two other gentlemen come into the glassed boardroom.

I am introduced, of course, everyone speaks English. Reinhold is the Corporate Controller for the hotel company and the other two, one is the Human Resources Director and the other guy I think was the VP of Operations for Russia and the CIS countries.

Then to my great surprise, he wanted to settle up the expenses right away. I found this to be most strange. Reinhold and I had previously agreed that my expenses were to be reimbursed in cash on my arrival to Vienna, but I wasn't really expecting him to make that the first thing on the menu.

It came across almost like he wanted to get rid of me immediately; "here is your money, now be on your way."

I would normally expect that to be taken care of after the interview not before, but I had no problem being given the cash right away. I just felt that it made for an uncomfortable interviewing atmosphere.

So there he was sitting beside me, he asked me to provide him with my airfare expense and other travel related expenses, cab fare, train ticket expenses and meals.

I did.

He then takes the receipts from me, adds things up and once again much to my surprise with the two other guys sitting there, he leaves the boardroom and comes back a minute or two later with an envelope and hands it to me, asking me to count the money.

In my experience, this was turning into the craziest interview so far I have ever had. Okay, so I take the money out of the envelope and proceed to count it. There were no coins; it was all U.S. dollars in various denominations, and it came to what my expenses were plus a few cents more since it was all bills. I thanked him and put the money away.

Now the interesting thing was that at this point I was still in the city. I still had to get back to Fischamend, and then take the train to Vienna the next day and from the train station back to the airport by cab.

He did not cover the return expenses for any of that part. I wasn't going to mention it. I was good. But knowing that; it put a sour taste in my mouth.

Reinhold had just now lent credence to the rumor about his hotel company being cheap. His decision not to cover my exit expenses confirmed that fact.

I am sitting at the boardroom table inside the glassed in room. The other two men were there, but I honestly cannot recall much interaction.

Reinhold did most if not all of the talking and interview process.

I don't recall a whole lot of what went down except for the most incredible part of the conversation and apparent decision making.

In fact, the decision in my mind had already been reached; I wasn't too interested; besides, these guys came off as interesting as wallpaper. The conversation somehow or other made its way into the area of spreadsheets."

Back in 1993, most of the business world was still heavily using the spreadsheet standard, which was Lotus 123. There were a couple of other competitors vying for dominance or at least an acceptable presence in the spreadsheet arena, one being Quattro Pro, and of course, there was the up and coming and fast growing Microsoft Excel.

I happened to be very comfortable and efficient with Lotus 123 as well as Quattro Pro. I considered myself almost expert level using Lotus 123. I had macros down pat and could fly around a spreadsheet developing a budget template almost with my eyes closed.

Reinhold got into this doctrine as I would describe it, about how their company has switched over recently to doing all of their accounting reports budgeting and maybe even their General Ledger using Microsoft Excel exclusively. I wasn't an "Excel" guy.

Reinhold went on about how it was imperative that I know Excel inside out because it would be my main tool as Controller at the hotel.

What was I to do? I knew that I could pick up excel without too much of a learning curve involved, but it did sound to me like he wasn't looking for anyone to be learning on the job. I was very confident in my abilities in delivering what was needed using the tools I knew how to use to the fullest and with expertise.

I responded. I said something along the lines of, "Reinhold, I understand you have decided to go with Excel company-wide. I have to tell you that this should have been discussed in our original telephone conversation because I am not an Excel user.

I am an expert with Lotus 123 and can design any spreadsheet you might need, but since you have migrated to Excel and you are looking for a Controller candidate who is an Excel user, well I have to tell you, I am not that person.

Although I can assure you, Reinhold, that for me to pick up Excel is not a big deal."

I could tell Reinhold was not happy. In fact, he didn't waste much time.

We talked a bit more, maybe five minutes or so about the hotel in Tiblisi but I could already tell that his decision had been made.

Now to this day, I honestly don't know if he had already made his decision before meeting me, maybe he heard something he didn't like about me, although I cannot imagine what that would be, or he really was just put off by me not being a Microsoft Excel user.

He did not say that he was passing on my candidacy; he said he would be in touch in a few days, but that was about it.

To this day, that was the craziest and most unprofessional interview I have ever had.

However, it did provide me with a wonderful experience and the opportunity to visit Fischamend and Vienna. All on Reinhold's dime except for my taxi and train fare back to the airport. Big deal! So, that was it.

***

My Mile High Club (sort of)

I don't know about you, but I definitely know about me. How profound is that? Well, probably not that profound but you probably know what I mean. Sometimes you might think that other people see things the same way you do, but they probably don't, but maybe at least they can kind of get what you mean. I think we are all capable of that.

I was single at the time. Come to think of it; I have been single for most of my adult life. I was married for a period, but not long. Longer than the 55 hours that Brittany Spears was married to Jason Alexander mind you, but not long, just over a year and a half. So for all intents and purposes, my adult life has been on the single side of things. Sure, I have had some relationships, some that I cherish forever with much love, but there were many in between periods of time when I was single and available and looking.

One of the most interesting and appealing circumstances that I always looked forward to, but only came to fruition one time, was to be fortunate enough in being seated on an airplane beside a knockout woman in my age group who happened to be single.

Now let me explain. What is so appealing about the "airplane environment"? Well, for me it was the perfect setup.

First, it was fate. There you were, or could be, the airline selects the seat for you, and they happen to seat you beside a goddess, and there isn't anything you or she can do about it. It's not like you're trying to weasel yourself in on a barstool to sit beside her. It's just the way things worked out, and hopefully, both you and she would be good with it; strangers that you both were, but welcoming the situation.

So, like I was saying in the beginning, I don't know about you, but I do know about me. For me, I have always wanted this to happen, even just once in my life, just one time please Lord!

Every time I have ever flown on a plane, my luck was pretty bad. I usually ended up being seated beside some doofus who annoyed the hell out of me by either coughing his lungs out or snoring or just being obnoxious. Other times it was the young mother with a baby, give me a break!

Where is the gorgeous super model looking babe who is going to take an interest in talking to me and giving me a "smile" (right, a smile) all the way to landing? So yeah, where is she?

Finally, my dream came true one day. All right, I understand that some of you reading this are women, that's good.

Remember now, that at the beginning of the book, I asked that you open your mind and try to see things through the eyes or understand through the mind of the opposite sex.

Well,…in this case, you might want to do that, or just sit back and enjoy how I look at things. I have to confess; I do not speak for all of my male comrades, but my thoughts are supported by a good dose of testosterone so that I will use as both qualification and excuse.

The training course I had attended in Walnut Creek California with my co-worker was over, and we were boarding the plane in San Francisco for the flight back to Miami. My co-worker was the front office manager for the hotel we worked at in Miami.

She was a nice lady, mother of two and the two of us from our hotel were to train on the new property management system being installed. So, that was completed, and I was looking forward to going back home and getting on with things.

At this particular time, it was already about six months or so since I was single again, after my divorce. You might know how things go; after a divorce, it takes a bit of time to get back into the swing of things, well by this time I was ready to get back onto that swing.

The flight to Miami from San Fran is a good five hours or so. I was ready to settle in. I was thirty-seven in 1988, and I had already flown a good number of times, probably fifty times at least.

So I was pretty familiar with the drill; seating arrangements, the service, even the types of airplanes and where the best seating was on different equipment (types of aircraft) from seaplanes to jumbos.

On this day I was a gentleman, and I offered to sit in the middle seat and let my co-worker take the isle. That kind hearted decision was to pay me back in a big way on this flight.

It was a Boeing 757 which had seat configuration of three on each side, with one aisle down the center.

I already knew I would suffer the trip back home crunched in the middle seat. I was hoping that the window seat would be occupied by someone who wasn't too obnoxious and didn't stink or wasn't coughing their lungs out or worse yet, with a crying, colic baby.

That by the way, was my fate on one flight over the Atlantic, and that was pure torture, this time, I could only hope to be saved from a repeat.

The announcement comes at the gate that our flight is ready for boarding. I need not get into the whole boarding process, but you know that it does take a while. It's a process, storing your bags, and things, finding your seat, some people sitting in the wrong seat then having to change, it can be somewhat of an ordeal, but eventually everyone takes off at the same time.

So I am making my way down the jetway slowly one step at a time, just shuffling along being patient and polite.

Finally getting onto the plane and now scoping out a good place for my carry-on, I start eyeing for my seat. I recall my seat was located over the wing of the plane and I look up ahead through the crowded aisle. It's not easy to see what is going on far ahead, but sometimes you can get a glimpse of what's what.

I ball-park my row from a few feet back, and I see that a big huge guy occupies the window seat.

At that point, my heart sinks. I figure he's already commandeered the armrest all to himself, and in reality, should be sitting in an aisle seat.

Well, as things would have it, ballparking my row from several rows back is not very accurate and by the time I do in fact reach my row I am elated!

Occupying the window seat is this gorgeous goddess of my dreams. I ask myself, "Lord is this it, is this redemption I have been seeking for all those other horrible seating experiences, can this be true?"

I cannot believe my luck! I am holding my boarding pass in my hand, and I look down at it again, just to make sure I'm not hallucinating, and yup; I will be sitting beside her for the next five hours! I'm standing in the aisle about to make my way to my seat, and look over at this goddess sitting by the window. She then looks up at me and offers a welcoming smile. Well, that smile did it. It was only a smile, but the smile was directed at me!

Now look, it's not like this was the first time a woman ever smiled at me. I was in the hotel business, and I had worked with and been around gorgeous women pretty much my entire career, from "Showgirls" in night clubs to gorgeous tourists sun tanning on Caribbean beaches. Not to mention the fact that my ex-wife was probably the most beautiful woman that I (and I do mean me) had ever laid my eyes on. That will be another "Life Tile," to be explored later.

At this point now I wasn't very aware of my co worker's actions, but she was getting her things ready to be stowed above. I eventually gave her a hand and took my seat in the middle. I had to say something, anything to the goddess after that flash of a welcoming smile, so I smiled back and said, hi, that seemed to do the trick. Simple and friendly, sometimes that's, all it takes.

My co-worker was still getting herself arranged and hadn't sat down yet, so for the first few moments, it was just me and the goddess. As I sat down and said "hi," she looked over still smiling and said, "looks like it's going to be a nice day for flying."

"It sure does," I said, with intentional double meaning implied, smiling back at her.

I wasted no time. I am usually pretty up front and make the most of things.

I said, "my name's Frank, nice to meet you." Then much to my surprise, and I wasn't sure why, maybe because I wasn't expecting, she said; "nice to meet you, my name is Claudine."

Claudine was the first "Claudine," I had ever met, and just hearing her say her name had me going. She was gorgeous, she was a fox, with piercing beautiful green eyes, and she had the greatest set of boobs I had ever seen in my entire life! Best of all she was in my age group, and I have no idea why but I just had this feeling that this flight was to be the best flight of my life.

Even now when I think back on it, hearing her say her name "Claudine" I cannot explain it, but it had sex written all over it.

Maybe it was the way in which she said her name or maybe because I had never met a Claudine before, but I have to tell you my mind went right to that place where sex is inevitable.

It had only been maybe a minute, but I was sitting there getting situated, dealing with my seat belt, strapping it on and just doing the getting ready thing, all the while thinking like I was hitting the jackpot.

"Okay well slow down here Frank," one might say. So, I will get to "the chase."

I would normally agree when hearing a story like this, wait, hold on, you're jumping the gun, you cannot assume things like that right away, that's not how things work, and life isn't like that. Frank, it's all in your head.

During the entire flight Claudine and I, we chit chat a little, nothing too serious, niceties and sometimes a few entertaining moments when we both laughed at something. The seating is close together, and there were a few times that our hands touched, but just touched, nothing more.

I, of course, wanted to take her hand and hold it and then lay a big wet one on her, but my imagination ran wild. Nothing I could do about it; that was just me and the way in which I'm wired I guess.

My co-worker is reading her book, consumed in it as far as I can tell, or maybe she's eavesdropping on my light conversation with Claudine. Regardless, I hadn't said much to my co-worker at all and more-less just left her alone to read her book. I am doing just fine with Claudine, but the initial air of excitement had calmed but make no mistake, she was everything and more than what I thought even before.

By the time the flight was to be over, we both knew enough about one another that the potential was there. The potential for what you ask? The potential for things to move forward. I said I would get to the chase.

We hear the captain come on the PA system to announce our approach to Miami International and for everyone to be buckled in. It will only be a few more minutes and my time with Claudine will be over, maybe over forever. Within the next couple of minutes, I would need to see if this is going to continue after we land. Is there hope in hell that it will?

I was debating with myself what to do. Of course, the answer is always no until you ask, right? So I had decided to ask, but I waited just a few moments.

Well, during that waiting period, of a few moments, Claudine spoke up.

Now the part that I found most unusual was that fact that she spoke up quite loudly. She took the words right out of my mouth!

It went like this, she stood up and said, "Frank, I would love to get together with you after the flight sometime and go for a drink on a nice patio, would that be okay with you? If you give me your number I could give you a call, and here's my number."

I could not believe my ears, and neither could my co-worker or the people I'm sure seated behind us and in front of us. Claudine had spoken loud enough for the big guy in the row ahead to turn around and look at her. I looked at him with a look on my face that said it all I'm sure.

Claudine and I exchanged phone numbers as we were deplaning and afterward we didn't say much if anything at all. We both just had smiles on our faces, and we parted ways. As for my co-worker; well she just looked at me, smiling shaking her head as we both walked down through the terminal to the baggage area. We caught separate cabs back to our homes, and that was the end of the flight from San Francisco to Miami.

Now the interesting part comes. Hold on; it has already been the most interesting and the most rewarding pleasant flight I have ever been on with great anticipation to come afterward and it wasn't long before it did.

At that time, I was living in North Miami Beach, just off Collins Avenue at The Plaza of The Americas. I had just moved in about four months prior. I loved living there. It was a condo complex, fully gated community with all the bells and whistles and I had lucked out big time since I rented a fully furnished absolutely beautiful condo for a song.

North Miami Beach along Collins Avenue was a great place to live. It was on the east side of the Intracoastal Waterway. The west side of the Intracoastal, well, to me that didn't have the appeal of the east side, the east side was where all the resorts and the beaches were.

It was my first real job in the USA, working as an assistant controller at the Castle Resort Hotel, a five hundred room famous resort on Miami Beach, just up from the Fountainbleau and Eden Roc which were both famous properties. Actually the resort I worked at was where the Playboy Club in Miami was located, but that was before my time.

I arrive back at my condo, and glad to be home. It was just late in the afternoon, being an early flight out of San Francisco, five hours plus the three-hour difference was eight hours plus a bit, so we were just winding down in the afternoon. My phone rings and I answer, it's Claudine. She would like to meet me for a drink later that evening and I of course accept!

It turns out that she lives not that far from where I am. She is up in Hallandale which is only about a fifteen-minute drive at the most, but we agree to meet closer to my place at "Shooters" which is on the Intra Coastal just right across from where I lived.

I was no more than maybe a five-minute drive from this very popular restaurant and bar.

This is definitely turning out to be my lucky day, and I am very excited to be meeting Claudine for a drink and maybe dinner at Shooters. Shooters, by the way, had a very extensive menu and is quite a large restaurant with both indoor and outdoor dining on the water's edge with a huge deck. It is a typical "Miami Vice" shooting location type of setting, very south Florida and very cool.

We waste no time at all. I am now of the opinion that we both wanted to meet equally as much after the plane flight, but it was looking to me like she was making all the moves and all I had to do was just be there.

I arrived at Shooters, parked my car and made my way inside.

Claudine was already there; I walked in through the front doors, and there she was waiting for me in the foyer, wearing heels and a mini skirt type of outfit. No stockings, she had the greatest tan and a big smile on her face as she saw me walk in. I couldn't have been more pleased.

So what happened on this day? I asked myself. Did I all of a sudden wake up in the morning with horseshoes up my rear end?

As I walked up to her, she reached out and grabbed onto my arm holding onto me, and we walked inside, through the restaurant and straight on out to the patio.

The restaurant hostess spotted us and showed us to a free table, just two chairs which I pulled close together, and we sat down waiting for the waitress to come and get our drink orders.

I am not going to get into what went down at Shooters, suffice to say we had a very very nice time and of course as we finished our meals and drinks, I think we had a mutual feeling that we had enough time together at this particular venue and it was time to leave.

I asked Claudine if she would like to come over to my condo which was just across from where we were. She happily agreed, saying, "absolutely."

By this time, I figured everything was as it should be. She was gorgeous and stunning, and she was with me! I wasn't going to question the gods that made this come about; it was just plain exciting, and this was my very first female encounter in Florida after having moved there a few months ago. I could not have imagined a better set of circumstances nor situation.

Apparently, she liked me, and I definitely liked her. Our conversation that late afternoon was cordial, we made some jokes, we laughed, it was good, and now we were both heading back to my place. I could only think of the inevitable during my drive back to my place.

Claudine was following me in her car, close behind me. Driving up to the gate, I informed the gate attendant that the car behind me was my guest and in she drove after me. We parked, and I then showed her around the area a little, the pool, the tennis courts before we headed up to my condo.

I was pretty darn excited, but I chose to be a gentleman and did not press, nor did I make any moves. We talked and drank some wine, and listened to music; the evening was going quite well, when Claudine then said, "well I guess I best be going, was really nice to meet you and spend the time with you, and thanks for dinner."

I thanked Claudine for coming over and getting to know her and told her that I would be happy to see her again. We hugged, and she left.

After having closed the door, I did feel disappointment. I remember that very well. I know it's not normal to meet someone so quickly, and right away you roll in the hay, it's just not the way to date if you want to make things last. But in fact, my little head was disappointed, whereas my thinking head said I did the right thing by not jumping all over her.

Now looking back on things and the events of that day and how things played out at my condo, I probably could have, and this is the reason why.

Claudine had been gone for about five minutes, no more than that. I remember I got myself another glass of wine when there was a knock on my door. That was strange, I walked up to the door, and looked through the peephole; it was Claudine.

My heart skipped a beat. I opened the door, and there she stood. I asked if she had forgotten something, and she said she hadn't but that her car has a low tire. I didn't have a chance to say much or anything as I recall, she continued saying that she didn't want to drive on a low tire and she would rather take care of it in the morning. Then she asked if it would it be all right with me if she spent the night. I said: "sure come on in," and poured her another glass of wine.

Claudine turned out to be my girlfriend for about three months. Our relationship ended as suddenly as it began.

I had immigrated to the USA, and I was living in the country on a "green card." I was legal, and I valued the privilege of being granted a green card. It was a cherished possession. I found out after three months time that Claudine was using cocaine on a regular basis.

Claudine was an addict. She never did it in front of me, and I had no idea until one day I found residue on my washroom counter. The minute I became aware of that, I knew I had to make the decision I did, and I ended our relationship.

I was not about to jeopardize my green card and my new life in the USA by getting caught up with someone involved in the use of cocaine and having it come back to me. The risk factor was too great.

After Claudine left my place, I no longer saw her again. As excited as I was about meeting her and the exciting times we had together, I was equally as sad and disappointed that she took advantage of me.

How did she take advantage you might ask? That would be by not being truthful about her drug habits and placing me into a potentially compromising situation.

But my desire to meet that hot, foxy lady had come true. Lesson learned: there is always more to things than what you see on the surface. You can never tell a book by its cover, so the saying goes, and how true that is.

My mile high club, was an on-going cocaine high, on a regular basis for Claudine, while I was unknowingly living the risk of being pulled over on any given day with a cocaine addict riding in my car and most likely carrying. A risk I could not afford.

Live and learn. Tread lightly when the rush to jump in beckons, I need to be more careful.

The next day I went to work without a girlfriend any longer, I knew things would change, and they surely did not too long afterward. I hope she is well wherever life took her.

***

Oh Island In the Sun

Oh my God, where do I even begin? No, really, I mean it. One of the main reasons I decided to get into the hotel business was because I had seen brochures of resorts in exotic places and that was just after I left high school and had no freaking idea as to what I was going to do for the rest of my life. There is a story about that as well. Another "life tile" to be revisited later.

This short story, however, does cover a bit of my time in the Bahamas. To begin with, you have to understand that I was thirty-one years of age. I had just finished working for the previous two years at one of the most luxurious and famous resorts in the world, being The Southampton Princess Hotel in Bermuda.

Now that too, those years being spent in Bermuda is another story all onto itself, but I wanted to mention that because it dovetails nicely into my transition to more years of island life, this time: in The Bahamas. The transition from Bermuda to The Bahamas wasn't as smooth as you might think.

There was some anxiety in between. After having completed my two years in Bermuda I was ready to move on and The

Bahamas was sounding very appealing. I had let my "headhunter" know that I had an interest in working in The Bahamas.

With me having Bermuda experience, I was a viable candidate for a good position to follow up with, in The Bahamas, and most likely a promotional opportunity or better said perhaps, a move up the ladder.

So, that is pretty much what happened, but getting there was somewhat of a nail-biter.

Having left Bermuda behind, and returning to Canada; I was unemployed, but I did have money, and I chose to go out west to Banff where I had good friends and where I could stay and pay rent living with one of my friends at his townhouse.

Doing that also allowed me to ski for most of the winter. I bought a ski pass and lived it up doing nothing but skiing every day when weather allowed, and having a grand old time in Banff, pretty much enjoying life.

Well, once again, I suppose I have horseshoes up my rear end, but that phone call from my headhunter did come.

I had secured an invitation for an interview to Nassau Paradise Island for Resorts International's Paradise Island Hotel and Casino for the position of assistant financial controller.

The resort was a twelve hundred room property with a twenty thousand square foot casino and a multitude of gourmet restaurants.

I flew down to The Bahamas (at their dime of course) and breezed through the interview, getting the job. Just like that!

One catch, though, I couldn't start until my work permit was approved and granted. I had to get back on the plane and fly back to Banff Alberta and wait, and wait, and wait.

I was offered the job in February. After I flew back to Banff and had skied my heart out for the remainder of the ski season, spring had arrived and I still had no work permit for The Bahamas.

As you can appreciate, I was becoming antsy. On top of that, I was running out of money. To take a job now in Banff just wasn't in my plan because I was waiting for the phone call to say that my work permit had been granted and I should get on a plane right away.

Well, April turned into May and still no word. I called my headhunter a few times and was told, "just to hang in, these things take time."

I waited, and now it was June. I was getting very nervous since I had no job and now almost completely out of money.

The situation was becoming desperate for me, the month was moving on, and I looked at my funds. I had enough money left in my bank account to book a flight to Toronto and have my dad pick me up at the airport.

That would mean having to stay with Mom and Dad at the age of thirty-one until I either found myself a job or the call from The Bahamas came. I had no choice in the matter; I had to go and stay with my mother and father back in Dundas Ontario.

When I boarded the Air Canada flight in Calgary after taking the bus from Banff, I had exactly fourteen, count'em, fourteen dollars to my name! In essence, I was virtually broke. Luckily my Dad was picking me up at the airport in Toronto. Otherwise, I would have had no way of getting to Dundas, which is just outside Hamilton Ontario, about 35 miles or so.

The plane arrives in Toronto, and my Dad is there to meet me. We leave the airport parking lot, and we talk some on the way back in the car. He could tell that I was depressed and I was down on my luck. He didn't say much, in fact, but he was being awfully nice to me and asked me to cheer up, saying that everything will work out just fine.

That actually did make me feel better because I was very uncomfortable becoming a burden on them. Truth be told, they both loved having me around. So then the real truth comes out after we arrive back at my parent's place. I bring in my luggage and greet my Mom. My Dad brings out the wine as he always did when I came home and as he is pouring me a glass, he tells me that he received a call earlier on in the day.

My Dad had gotten the call that I was waiting for all along for the past several months. My Dad tells me that the hotel in The Bahamas, had tried calling me in Banff earlier that day, but I had already left for the airport.

My boss to be in The Bahamas had my contact number in Dundas for my parents and he called my Dad my work permit had been approved, and I was to go pick it up in the next day or two at the headhunter's offices in Toronto. After that, I was to book my flight to The Bahamas, which the company would arrange for me.

Well, I was elated! No wonder my Dad was telling me not to fret, he had kept it under his hat till I was home with my Mom and we were all together.

My brother; Jim, also had come over later that day who happened to live just a few minutes away. So, that was how that all went down.

Apparently, my work permit took some time due to additional arrangements having to be made for it to be approved. In other words, someone in the work permit issuing office needed some more pocket money before he or she would sign and approve it. That's how things were done in The Bahamas.

Now, getting back to me being thirty-one years of age and single, and on top of that, landing a job that placed me smack in the middle of one of the world's most lavish, exotic, exciting and most desirable of resort destinations. Well, I ask you.. how much better could life actually be for a young man at a perfect point in his career to take advantage of everything that came in this package? Answer: not much. It was as good as it gets in my mind. I had it made!

I did not need to die to go to heaven; I truly believed that I had arrived in heaven when I finally arrived on Paradise Island Nassau Bahamas as Assistant Hotel

Controller for the most prestigious resort in the Caribbean.

The flight over from Miami to Nassau, how awesome was that? It was incredibly awesome.

I had landed in Miami the day before actually. The corporate head offices for the Bahamas property was in Miami, and I was asked to drop in there to meet some people before flying over to the island the following day. Well, at this point I wasn't aware of the fact that the company also operated the oldest airlines in the United States, being Chalks Airlines, which was a seaplane airline service.

Chalks had a fleet of Turbo Mallard Sea Planes and the Grumman Albatross operating out of the Chalk's seaplane terminal on Watson Island on Biscayne Bay.

The following day, I was given a ride by one of the people I met at the corporate offices to the seaplane terminal to catch my flight over to Nassau on the seaplane.

How gorgeous, and what an experience that was! The Grumman Albatross taxied; rolling down the ramp into the main shipping channel waters of Biscayne Bay.

It was very exciting to hear the sound and feel the power of the turbines rotating the propellers generating the speed to send us skimming over the watertop faster and faster lifting us up and over the line of cruise ships lining the Miami terminal. Quite breathtaking, and I was on my way!

Then heading out over the blue Atlantic, rising higher but nowhere near the altitude of jets; provided such a fantastic close view of the ocean below. The ship traffic, and the pleasure boat traffic, it truly was exciting. Once having flown over the Gulf Stream, which was 45 miles of dark blue, deep water, it gradually gave way to brighter and lighter colors of the ocean, and soon the famous yellow sandbars of the Bahamas came into view. How beautiful was that! I had already been to Bermuda, but this, this was like flying into never-never land.

The ocean and the scene unfolding beneath me were absolutely breathtaking and the closer we flew to Nassau the more beautiful things became. I was so happy at this point that I cannot put it into words, although I am doing my best to do so here.

The seaplane terminal was located on Paradise Island. The seaplane came in for a landing on the water in Nassau Harbour. For the water landing approach, it flew level for a while with the upper decks of the cruise ships lining the cruise ship terminal in

Nassau, another very picturesque scene. Then, touching down onto the water, felt like the plane was being pushed along on a bed of marshmallows.

You see, the seaplane has no pontoons, like most seaplanes. It actually landed on its belly.

Underneath the wings, are two outriggers on either side, but you feel the landing on the water transfer directly to your feet. As you taxi along the harbor; the water is right outside your waterproof windows, and sometimes the waterline is halfway up your line of sight looking out, so it's a very unusual experience, most enjoyable and very memorable.

The plane then makes it to the terminal and the wheels come out from the sides of the fuselage and up it comes onto the ramp leading to the terminal.

You see, the plane is amphibious, could land and take off from either the ground or the water. We arrive, check in with customs and someone from the hotel is there to meet me.

"Welcome to The Bahamas Mr. Julius." Damn I felt good!

The funny thing was that when I had boarded the plane in Canada to fly to Miami, I was still technically broke. My last fourteen dollars that I arrived with from Banff, well I had burned through that, and I had to borrow money from my parents to get me through at least till I landed in The Bahamas.

As things turned out, I spent no money even while in Florida for that one day. A company employee had picked me up at the airport when I arrived in Miami, my hotel was covered by the company as well, and the drive over to the seaplane terminal the next morning was provided by a company employee as well. My meals the day before were also taken care of; having gone out for dinner with a few company people the night before.

So here I was on Paradise Island with not having spent a penny getting there. Soon I would be earning my first pay cheque, and I could repay my parents. I wasn't thinking about any of that, but now looking back on things, yeah, I hadn't spent a dime getting there; things were really looking up!

◇◇◇

No, I was not yet at this point of my career a member of the executive committee. I still had a rung or two of the ladder to climb but having the position I did, still afforded me with almost the same privileges and perks that the big wigs enjoyed.

For housing, I had my own executive suite in the hotel on the 7th floor overlooking the turquoise waters of the Atlantic, looking out onto the beach on Paradise Island. That is where I lived. Every night I could fall asleep listening to the soothing, steady rhythm sound of the Atlantic Ocean brushing the beach, just outside my window, seven floors below.

I had free food and beverages, whatever I wanted at any time, including room service as well as unlimited food and beverage requisitions to my suite from the hotel storerooms. On top of that, I had full use of all the restaurants in the resort, and everything was on my in-house duty meal account. That too, was a part of my perks or employee benefits, but because there is no income tax in The Bahamas, even my employee benefits were not taxable, there was not tax. Now isn't that just more icing on this cake?

The only thing I did have to cover out of my own pocket was fifty percent of the charges that would be incurred by anyone I had visiting. That would include my family or my friends. So, there I was, working in a fabulous resort and getting paid every two weeks and not having to spend a dime to do anything. Does it get any better than this?

My job actually was a great deal of fun. Okay, it was accounting and internal controls, and there was a lot of that to be done. Not going to get into all of that now, but I did have a considerable amount of responsibility that made my days fly by quickly. Talking about the days flying by quickly, at the end of each working day, which was around 5:30 or so, my boss and some of the other hotel execs; we would all get together for a drink in the lobby lounge.

It was like a five o'clock club. Sometimes the General Manager of the hotel would join us; sometimes it would be three or four of us, sometimes it would be a much larger group including some of the Bahamian managers. There were only six expatriates working at the resort in management, three Canadians, one Brit, one American and a Swiss. My boss was one of the other Canadians on staff; he was a pretty decent guy, after all, he did hire me.

I believe it was my first week working at the resort that I faced my first test so to speak.

One would think that the sort of job I had, being in accounting and controls would more-less exclude me from interaction with hotel guests. Well, nothing could be farther from the truth.

The position of the assistant controller or that of controller involves a great deal of guest interaction, especially in a resort that happens to have a casino attached to it. All the front desk procedures in as far as accounting for guest charges, room rates, and anything and everything that has to do with money, comes under the authority and responsibility of the hotel controller or assistant controller.

In fact, I would venture to say, that the assistant controller had more interaction with a hotel guest on a daily basis than did the hotel General Manager. The GM was usually consumed in endless meetings in either hotel operations related matters or sales and marketing meetings with the resort's department heads and executive committee.

Anyway, these two individuals are directly responsible for ensuring the collection of all accounts and looking after the assets of the property. That would include, inventories in food and beverages, all the housekeeping supplies, all the furniture in the hotel and the list goes on and on for miles.

So, also included in that list was the area of credit management, and I was the de facto Credit Manager.

Back in the early 1980's of course we did not yet have the internet. Obtaining approval on credit cards required an actual phone call to the credit card issuing company and obtaining an approval code, which then was written onto the credit card voucher, and submitted to the credit card company for payment to the hotel.

Well no matter how closely the establishment monitored guest accounts, it was inevitable that some guest accounts exceeded the credit card's limit. That then resulted in the guest having to settle his hotel bill with cash or traveller's cheques (which by the way were very popular back then) or perhaps a different credit card.

Many a time, hotel guests would spend their last dollar at the casino's tables hoping to regain their losses; only to be totally cleaned out and unable to pay their hotel bill the following morning when checking out.

The resort was vigilant in collecting what it was owed. Each and every day I reviewed all guest accounts that carried a balance higher than two hundred dollars.

To put you into the picture, it didn't take much; even back in 1983 to run up a hotel bill of incidental charges into the hundreds of dollars. One room service order could easily run into a couple of hundred bucks.

For example, if you were feeling lucky the night before checkout, you might think of living things up a little and throw caution to the wind; pick up the phone in your hotel room and ring room service.

Your order might run something like this; two shrimp cocktails as appetizers, one lobster thermidor, one steak Diane, and say a couple glasses of wine and bingo; there you go, two hundred dollars with tip.

On top of that, you might have a few drinks in the bar which you also sign to your room account, and then you might have a sudden urge to make an overseas long distance call back to St. Louis and twenty minutes later you just created an additional fifty dollar charge to your hotel bill. You and your wife or girlfriend, or fiancé have thrown caution to the wind being your last night in The Bahamas.

The following morning you both decide to grab a little beach time before checking out at noon heading for the airport.

Much to your surprise, however, when you try accessing your room, you find that your key no longer works and you are forced to come to the front desk and complain that you've been locked out and you still need to pack and make your way to the airport in time to catch your flight.

This was usually where I came into the picture. The resort enforced a strict policy of all accounts being cleared and paid in full upon departure. Guests with high balances had their credit cards pre-authorized to facilitate a smooth and trouble free checkout. However, when the credit card company declined authorization, the resort had to take action.

I had only been in my new job for about a week or so, but in my first week I was already subject to baptism by fire so to speak. You see, the hotel wasn't in a position to give the negligent hotel guest, benefit of the doubt. As history proved, in most cases, the hotel would never see the money once the guest had flown back to the USA.

Normally we would call the room and ask that the registered guest come to the front desk But by this point, the hotel was already in a potential loss position, and very probable that the guest was thinking of bypassing the checkout procedure and just catching a cab to the airport.

If the guest responded by saying something along the lines of, "we will be down in a while," or "we will be checking out soon," my response would be; "no sir, you are required to settle your account immediately. A hotel security officer will escort you to the front desk."

With the situation having reached this point already, a hotel security officer had already been dispatched outside the guest's room to prevent the guest from just skipping out on the bill or waiting for my instructions on his radio, at which time a knock would come on the guest's door.

If the guest came willingly, then no problem, security would not be in their face. However, if upon calling the room and there was no answer, security would enter the room and determine whether the guest had already skipped. If luggage was still present, it was a safe bet to assume the guest was still at the beach. The security officer would then double lock the guest's room to prevent re-entry.

Naturally not being able to get back into their room, the guest would be forced to come to the front desk at which time they would see me.

Now remember, in most cases, the guest or guests, and it was usually a man and a woman, would be making their way back from the beach or the pool.

We are in The Bahamas; the resort is a beachside location, and once you are settled in, it's very common and normal to for you to be walking throughout the public areas in beachwear. Women in skimpy little bikinis, some women would wear sarongs, some not. So whether the guest was wearing a suit or standing there in a speedo; it mattered not.

Either way, I would ask them to my office. So, now picture yourself if you were the one not being able to pay your bill, sitting in front of someone who is behind his desk, and flanking you are two burly security guards and the chief of hotel security, while you and your girlfriend or wife, are sitting there pretty much naked. It does paint a picture of vulnerability does it not?

I would say that in about 90 percent of the cases, when the guest's credit card had been declined, and the guest had not arranged to pay down his account, the guest was at a loss and wasn't able to come up with the cash.

What to do, what to do? The solution was simple enough.

It went something like this; "Mr. Jones, seeing that you are unable to settle your account, we are required to hold an equivalent value of your personal property until such time that you send us payment."

At this point now, both the man and woman would just be sitting there totally bewildered that I could do this.

I recall a couple of times when the gentleman claimed he was a lawyer.

"You can't do that, I will have you and your company sued so fast you won't know what hit you!"

At that point my response was usually something along these lines.

"Well Mr. Jones, I appreciate that you are a lawyer back in the United States, but here in the Bahamas you are just another tourist, and your law license doesn't extend here into The Bahamas. This is a different country with different laws. Now if you prefer not to cooperate in the manner I have laid out, then Mr. Munnings our Chief of Security will be happy to have you taken to jail and spend the night there until you make arrangements for discharge."

Having said that, the woman usually started crying, not to belittle women in any way, just is what happened.

Sometimes arguments between the two would ensue.

"I told you not to order all that food, and why did you have to gamble away our last dollar at the table?"

That sort of thing and it became ugly. When arguments started there in front of me that is when things became dicey.

In most cases, I would end up holding back their luggage and all of the content. In essence, they would depart the hotel without any of their belongings. But as simple as that sounds, they would start arguing about what to leave behind. Sometimes they would insist on taking their luggage and belongings back home with them.

If they chose to take their luggage, then other forms of collateral had to be secured.

I hate to admit it, but I did, in fact, withhold wedding bands and engagement rings, with crocodile tears streaming down one woman's face as she slid her ring off her finger, handing it over to me. "Nice guy I am eh?"

Another time, I held back a gentleman's Rolex Oyster wristwatch and a woman's diamond ring. Over a period of four years at the resort, I believe I ran through this scenario at least a dozen times, and by that time I had a reputation as a ruthless collector!

Arrangements were then made, with paperwork filled out before the guests' departure. Everything was made whole again with items returned to the guest once payment via western union was received.

That was how things got done in The Bahamas. I am sure there were some couples who had a very miserable flight back home.

Now one little interesting aspect to this scenario and one situation that happened which sort of made me laugh and cringe at the same time but then again thinking back on how things go, well…let me explain. I got a laugh out of it and maybe you will too. Here's what I'm talking about.

I suppose I had been there on Paradise Island for a couple of months. I was at this stage pretty much involved in all aspects of the resort's operations, from a controls and assets point of view. I'll get back into a credit issue once again, with a guest not being able to pay her hotel bill on a timely basis; that is how this incident unfolds.

The thing is that you could actually stay at the hotel without a credit card on file so long as you had made a large enough deposit in cash ahead of time to carry you through for at least two consecutive days. That would require you to pay up front two nights' room rate as well as put down equivalent amount for incidental charges. These types of "cash" guests, I had to keep a very close eye on and as soon as their charges to their room account was starting to eat up their credit balance, I would make sure to ask for an additional deposit up front to cover for the next couple of days.

Well, this was a resort with a casino attached to it. The most luxurious and lavish in all the Bahamas, with a cabaret theater and several lounges, discos, bars and several gourmet restaurants adjacent to the casino. One very famous restaurant featured in the James Bond movie, Thunderball, called Café Martinique. A lot of that James Bond movie filmed right there on Paradise Island; the resort was very popular, a destination resort.

Now the thing with the casino environment in The Bahamas was that it was a different animal from your Vegas casinos.

In The Bahamas, most of the casino activity was taking place in the evening, after beach time; unless it was raining of course, and then the casinos would be jam-packed all day long. But in the evening, pretty much all of the hotel guests headed to the restaurants, and after that, it was time to enjoy the casino.

Before you ventured out for the evening, it was traditional to be dressed to the nines. Ladies in elegant evening gowns, the gentleman in smart outfits and usually a sports jacket and never blue jeans, just wasn't done, dress pants were in order.

I was amazed after a couple of months on the job, how elegant the tourists were dressing, and let me tell you there were lots and lots of gorgeous women around all of the time.

I had particularly become aware of a few women I happen to come across on a regular basis. I would see them from time to time in the casino, then not for a week, then see them again for another week.

I wasn't sure if they were tourists, or lived on the island and liked to gamble or what, but they became familiar faces. Women in their early to mid-twenties, but absolute knockouts.

They, in turn, would see me as well, and it got to the point where familiarity with one another became apparent, and we started acknowledging one another after a while, just more-less nodding in passing.

Well, I thought I knew pretty much all there was to know about the hotel by that time regarding our room rates, perks, guests' privileges, etcetera. Our rate structure at the resort allowed for discounted rooms for certain individuals such as travel agents, diplomats, tour consolidator, guests of management and casino high rollers. The high rollers' rooms, of course, were picked up by the casino. The casino although attached to the hotel, was a separate business. The casino actually paid for the hotel rooms occupied by its high roller gamblers. Although the hotel suite or room was complimentary to the player, the hotel still received revenue from the casino for that room.

So one morning upon reviewing guest accounts for high balances, I came across a cash paying guest, whose account was way over limit, and no longer in credit balance by at least a couple of hundred dollars. Well, I did what I normally would do and called the room asking the guest to come down and settle the account.

It went something like this. "Yes, good morning Ms. Freeman, this is Frank Julius, I am the credit manager for the hotel. The reason I am calling is to ask you to make a further deposit to your account in order to settle the outstanding debit balance and bring your account up to date with a deposit covering your next two days. If you would be kind enough to see me the front desk, I will be happy to handle your payment process personally."

I finished waiting for a response, and it came.

"Oh, Mr. Julius, I will be right down to take care of things."

So good to go right? I am waiting for Ms. Freeman to come down and within about a minute or so the front desk clerk calls me to inform me that there is a Ms. Freeman asking for me. My office was situated just behind the front desk, I then walk out from behind the front desk, and much to my surprise it's one of those lovely ladies that I have been nodding to for the past few weeks. So you know where this is going right?

Well, I don't have to tell you, Ms. Freeman was hot. Oh, my God, she was hot! Wait, did I say she was hot?

She was about my height 5'7", she was dressed wearing a mini skirt, and she had legs that ZZ Top's song, entitled; "Legs" was all about. She had a big smile on her face, and a demeanor oozing of sex and erotica. To be honest, I hadn't realized just how hot she really was in all the passing times I had been nodding to her, but now standing close, well wow!

She most definitely recognized me, and although I had seen her for a better part of at least a month and a half, I am sure she knew that I was in some kind of management capacity at the resort, but I don't believe she actually knew what I did at the hotel. She was about to find out and to balance the equation I too was about to find out about her.

I saw right away that she recognized me, she had this look on her face that transmitted a sense of relief as she smiled at me when I came to introduce myself and then gave me a slight but most definite wink.

Well, she was a hooker for God's sake, but with the personality of a talk show host, bubbly, most affable and gregarious, and on top of it, she seemed very confident about herself. So although I figured she was a hooker, I liked her because of her friendly outgoing cooperative personality, and I did not consider her to be a flight risk on her account.

I approached her situation as I would with any other guest, and invited her to my office. By this time, employees located in my area of the hotel found it commonplace for me to have guests, travel agents, and tour operators, in my office discussing their accounts and needs.

My office by the way, as I mentioned was located behind the front desk. First, a guest had to be directed to a door located in the lobby that leads you down a hallway taking you past the General Manager's office, the GM's secretary, then past my boss's office and his secretary, that then further opened up into the main accounting office.

The main office was a large room with desks situated around the room with employees doing various jobs pertaining to accounting, and my office was a walk across the open accounting office leading to a door to my private office.

So essentially, everyone from the GM's secretary to my boss, all knew by the time I reached my office with Ms. Freeman, that I was taking a lady of the evening into my private office. I sensed they all knew, and I kind of laughed about it all, but I had to do what I had to do.

I closed my office door, and I did that more for effect than anything else. Plus it gave the employees something to talk about; I was having a little fun, and I think they like it.

"Ms. Freeman, thank you for coming down to see me, I need to have you pay your account in full and provide the additional deposit to carry you through to your indicated departure date."

She was sharp.

"Frank, it's so nice to meet you, I see you all the time walking through the casino." She picked up on my name, and that impressed me. I later learned that most of the "girls" working the casino were very sharp, well educated and could engage you in the conversation of your choosing. Most of these girls were university students working their way through school. Ok, I'm not kidding! They were!

We cut to the chase.

"How much is my bill and what does the total come to for the balance of my stay?" I don't remember how much it was, but it was substantial.

She then said. "Well I don't have the money right now, but if you give me a just a couple of hours, I can have it. I'll get in touch with my girlfriend Wendy, and we'll have the money for you in a jiffy. I'm sure you've seen Wendy around the casino, I can introduce you to her if you like." Well, I didn't need any introductions to Wendy, but I think I knew who she was talking about. So what Ms. Freeman wanted was for me to give her a couple of hours so she could turn a couple of tricks with her friend Wendy and have the money to pay her bill for the remainder of her stay.

It was very obvious to me, that when she said that she would have her friend help her out, well they were going to offer their services on a tag-team basis, two for one deal, a double header, a quick money maker, call it what you will.

I was certain that they'd have no problem finding a client in the casino who would be more than happy to entertain their offer. I'm not sure, but I would guess that would command about three to five hundred dollars per session of entertainment for a "double header."

I agreed and gave her three hours to come up with the money, or her room would be double locked and her belongings confiscated as per hotel policy on non-settled accounts.

Ms. Freeman was back two hours later with the money and took care of her account. I will never forget, after my office conversation with Ms. Freeman, as she was leaving my office, she couldn't help but adjust her dress as she walked through the accounting office implying that something had gone on between us behind closed doors.

I was standing in my doorway watching her leave, and I just laughed it off, and so did the staff.

By this time, word had gotten around that I had one of the "girls" in my office to settle her account.

The resort's front office manager, "Peter," was a local Bahamian, a great guy, he was well connected in many ways, at the resort and of course on the island. His personality was tailor-made for the hotel business. He was a real pleasure to deal with; someone who knew how to make you feel at home, a great Bahamian. I asked Peter to my office and had him fill me in on what was what with these girls.

Even to this day, I do not know if the senior management of the resort and the casino ever discussed the situation concerning prostitutes, ladies of the evening, hookers, call girls, call them what you will, or ever developed a policy regarding this activity. After Peter had come to see me, I then understood that the resort offered a discounted room rate to these gals.

Now the official reason for extending a discounted room rate was to acknowledge their loyalty to the property and their frequency of stay, which would not be out of the ordinary, but these gals were anything but ordinary.

In addition, I never did find out, not that I pursued the topic, but I don't know if anyone from the resort was, how should I say, "in charge" of which girls were allowed to stay and which ones were discouraged and asked to move on. It seemed to me that the casino tolerated their presence so long as they remained high class and caused no problems.

So there you have it, and over the coming months and years I got to know most of the women in this line of work who frequented the casino and stayed at the resort. They paid their bills on time, and never a problem.

I recall another funny incident though. I had been working there for about six months. My brother Jim came to visit me for the Christmas holidays from Canada. That first afternoon, we were walking around the casino and the resort. I was showing my brother around the property, the grand tour, getting him familiar with the layout; where everything was and a feel for the place.

Well, during my tour showing my brother around, we had come across a couple of these women who already knew me and smiled at me big time and my brother as well. He was a good looking twenty-seven-year-old, six feet tall; the women went for him big time.

But anyway, we're walking around the casino and a couple of these girls are eyeing him big time. My brother's eyes are popping out of his sockets as he says to me, "wow bro, there are some really hot women here, I notice some of them were smiling at me big time." I replied, hey cool your jets bro, those women are hookers."

"Well, no wonder they were coming on to me like that."

I just laughed it off. Trust me; my bro had no problems meeting women at the resort while visiting me, there were plenty of single women on Paradise Island at any time of the year, he didn't need any of the working girls' favors, but it was funny nevertheless.

Working at a famous resort in the Bahamas had its benefits. I got to meet some very interesting people who vacationed there.

I met movie stars and politicians, sports celebrities and local dignitaries over the years and I was in a position to invite close friends from Canada as well as my family to visit, stay for free and make their time fun and enjoyable.

One night I met my future wife to be who happened to be on vacation, staying at the resort, and a year afterward I left the

Bahamas, immigrating to the USA to be with my wife and lead a life in Florida.

I have lots of stories about life in The Bahamas, but I thought the one about the ladies of the night was an interesting one.

I learned not to be quick in judging people. Sometimes my preconceived notions could be totally wrong and not serve me well. I think in The Bahamas 1 learned more than ever just to live and let live.

There was an entirely different set of values and way of life. Some people kept telling me that it wasn't reality, and everyone had to return to their real daily lives back home after the vacation was over. Well for me, it was a reality, it was my reality, and I dealt with it every day the best I could.

You would think that living in Paradise was a piece of cake. It did, in fact, have its own set of challenges. The resort was, make no mistake a multi-million dollar business that had to operate efficiently.

Feeding several thousand people each day, and looking after their safety, well-being, comfort, and entertainment hopefully well enough to ensure their future return. Bringing them back to the island; back to The Bahamas and keeping everyone on the island employed and happy. They say, "it's better in The Bahamas," and I couldn't agree more.

***

Welcome to Russia (NOT!)

One of the most incredible events in recent history was when President Reagan went to Berlin, and on that visit of June 12th, 1987 said, "Mr. Gorbachev, tear down this wall."

Those courageous and forever echoing words in history spoken by President Reagan planted the seeds that then grew into a movement with a domino effect over the next three years that lead to the fall of the Berlin Wall and the breakup of the Soviet Union.

Many events had taken place that eventually led to the fall of communism in 1991 with the ouster of President Gorbachev.

Ironic as it was; Hungary had a very significant role in stoking the locomotive into action that would pull through the events of history bringing freedom to all of Eastern Europe and disappearance of the 'iron curtain."

These historical events then allowed me to accept a consulting position in Moscow Russia, in 1992 just a few months after Mr. Yeltsin stood atop a Soviet tank and declared the end of Communism and the birth of free enterprise, along with newly realized widespread freedom for all Russians.

The winds of change blew briskly in these early years of upheaval, formation, development, and change. I played my part in this period of history. I made a small but indeed a contribution to it all.

In some ways very ironic and other ways very natural. The world was changing, and I took the ride on the train of change that rolled along the tracks of history heading for the station of inevitability.

One important factor in these life tiles is one tile that stands out, that tile, bearing my place of birth being Budapest Hungary in 1951, and now I was about to enter the former Soviet Union.

It was all because of the former Soviet Union that our family was forced to flee my birth land and seek freedom in the West. Ironic how I now found myself going to the place that ran us out of our own country.

Yes indeed, the winds of change had blown me back over the Atlantic, and now I was about to deplane and enter Russia.

I first personally felt the breeze of these winds of change only a month before my arrival in Moscow. I had been working a project in the British Virgin Islands on beautiful Virgin Gorda in the Caribbean. That project was coming to an end. While working at Little Dix Bay resort, I received a phone call from my headhunter.

He informed me that he had a short term opportunity for a consultant to do a general accounting project for just three months in Moscow. It was for a newly opened property run by a Canadian management company and would I be interested in going to Moscow.

It was a most unexpected and very out of the ordinary opportunity. I first resisted and told the headhunter that I wasn't really interested.

I hadn't exactly shut the door tight on that phone call having said that I wasn't "really" interested.

So the next day, another call comes but this time it's from the hotel in Moscow and its the Director of Finance calling. He then tells me that he will be in Toronto in two weeks time and would like to meet me at the head-hunter's offices, and he'd like to make arrangements for me to fly up for the day.

Okay, so all expenses to be paid and it dovetailed nicely with the completion of my project there in Virgin Gorda.

I hadn't seen my mother or brother for a while anyway, so it was a freebie trip to Canada, so it was no skin off my nose, why not go? I accepted his offer and went to Toronto, but I drove instead of fly. That was fine by him as well, still picked up the tab.

Upon meeting with him two weeks later, he had a good offer on the table for me, and I decided to take him up on this adventure and go to Russia.

Even though I decided to go, I still had reservations about it all. Somehow things still did not sit well with me in going to the nation that held Hungary under its thumb since the Second World War in virtual slavery.

I know that the term "slavery" is a bit harsh, but in essence when your country is surrounded by land mines and barbed wire, and you are not free to leave or to travel, well that in my book is slavery. The official line, of course, was that it was there to keep people out! Well, give me a freaking break! How many times in history has the world ever heard of someone making a run for it, to East Germany or making a break for the Hungarian border from Austria? It was totally absurd, but that was the official line from the masterminds of the government controlling Hungary, a puppet government to be sure, but still; the government of Hungary.

The most incredible thing then came about in 1989 with the changes taking place all across Europe.

Hungary did in fact stand up to its Russian Overlords once again in 1989. That became known as "the peaceful revolution of 89," when Hungary took down the barbed wire near the Hungarian border town of Sopron, and an influx of ten thousand East Germans poured into Hungary, actually making a run for the Hungarian border.

Who'd have ever thought that to come about! But it did..people actually escaping into Hungary! And from there on to further freedom in Austria and western Europe.

It was Hungary's Foreign Minister: Horn, who took the initial steps that gave life to the "peaceful revolution" of 1989 and got the whole ball rolling. The following momentum could not be stopped.

The Russian troops stationed in Hungary took no action. By that time Gorbachev had seen the writing on the wall very clearly and soon after that, the Berlin Wall came down. The Russian troops eventually withdrew from Hungary; the Soviet Union was no more, and I was on my way to Russia.

I had done a lot of flying already by 1992, but I had yet to experience a flight to

Europe. There were many times I wanted to go, especially a trip to Budapest but it never came about.

My flight to Moscow was to be on KLM airlines; a great airline, by the way, with a short stop and transfer of planes at Schiphol Airport in Amsterdam and then continuing to Moscow.

I have to tell you about my arrival in Russia. The flight from Amsterdam to Moscow was totally different than the leg from Toronto to The Netherlands.

The plane was now filled with Russian speaking passengers.

I tried listening for English voices, but I did not hear anyone speaking English. I already felt like I was in Russia.

Also, I felt the atmosphere to be different on the plane. There was a certain sense of, how should I describe it? "Gloom," I think is a fitting word. That air of gloom permeated through the aircraft, and it made me feel uneasy. Of course, it was all in my head and all, but sitting there I remembered back in the late 50's and early 60's we in Canada even were subjected to the ridiculous public safety commercials on TV called "Duck and Cover." You might remember those stupid and idiotic commercials if you are old enough. For the younger readers; the

TV commercial on our black and white RCA TV with a 12 station channel selection dial, would instruct us to duck under a table and cover our heads in case of a nuclear attack from Russia. Just totally insane, but yeah, that aired for months on end.

I clearly remember the commercial even after thirty-five years, showing a classroom of elementary students and suddenly the air raid sirens coming to life, then all the children in the classroom suddenly jumping off their chairs and ducking underneath their classroom desks, covering their heads with their hands to protect themselves from nuclear annihilation.

Now that truly was absurd, and yet, that was how things were presented to the masses back in those days; totally insane.

I wondered how much more of our North American indoctrination to the Russian Bear was just a bunch of made up political malarkey. Well, I was about to find out for myself.

It was late September when the plane landed at Moscow's Sheremetyevo Airport. There was no Jetway; it was a drive-up staircase to the jet and then a walk to the terminal and Russian customs. I figured there would be two pieces of I.D. that I would be asked to produce, maybe three. Upon entering the airport and the Customs area, I can assure you that the welcome committee was not on hand that day.

That feeling and atmosphere of "gloom" was even more prominent as we cued in line waiting our turn for the Customs Officer.

Things were very "military-like." There were soldiers around with guns of course, and there wasn't a smile to be found anywhere.

Then, standing in line, it was my turn. I walked up to a kiosk type of little cubicle where the Customs Officer was inside, behind a window with a hole in it and a spot for me to slide my documents through.

When I walked up and stood in front of the kiosk looking at the nice Customs Officer things weren't getting any better or friendlier. There was a pool hall sort of light and shade shining down on me and onto my documents that I placed on the sill in front of him and me.

I had a Canadian Passport; I placed it on the sill. The Customs Officer whom I took to being just another soldier, he was certainly wearing a similar uniform as the soldiers standing around with guns, sized me up and compared my passport photo to my face. He then asked very emphatically in English with a heavy Russian accent, "where you live?"

This is where I figured I would be maybe a little "fucked."

God, I wished right then and there, that I would have just stayed back home and not come to Russia.

So I answered him. "I live in Florida United States."

He was very intimidating. He then asked. "You live United States, your passport Canada, where you born?" Now I figured my hole had been dug even deeper; now I am more than a "little fucked."

I answered, "Budapest Hungary."

At this point, I thought my goose was cooked, and he was going to have me follow him to a little room for further interrogation, thinking I was maybe a spy."

Hey, don't laugh; this sort of shit went through my mind in a very serious way. I had heard stories, and besides, I had watched James Bond's "From Russia with Love" and all other James Bond movies my whole life! Cut me some slack. Okay, I was only joking about the James Bond movies, but still!

I was pretty concerned at this point standing in front of Mr. Russia not feeling real comfortable about anything at that point. You see, my father had been a freedom fighter in Hungary in 1956 and had blown up Russian tanks and killed Russian soldiers. He was being hunted by the Russian KGB and Hungarian AVO back in that time, which caused us to flee Hungary. Now all of that history, was suddenly front and center in my mind, thinking they haven't forgotten about that shit.

My father never had the desire to return to Hungary after escaping. He always told us, that he would be taking a chance and it was possible that if he went back to Hungary for a visit, he might not be allowed to return to Canada, so he never went, didn't want to take the chance.

But the winds of change were blowing, right? I was standing there, fretting over my future. But the winds of change were blowing, I kept telling myself, now have me just breeze through this, and I'll be good to go.

The Customs Officer then pressed some more and asked, "your reason for coming to Russia?"

I pushed my work permit forward underneath the gap at the bottom of the window. He took it and read it. He then looked up at me, and pointed at me, and said, "you wait here," and he left with all my docs and passport.

I was getting antsy. Stood there in front of the kiosk now with nobody in it, and looked behind me at the line still not depleted, it was long enough, and I could tell people behind me were getting sort of concerned that I was taking so damn long. The line I was in was for non-residents, so I'm sure the foreigners still to come after me weren't feeling the warm and fuzzies seeing what I was going through.

So I waited some more. It seemed like forever but it was probably only five minutes or so before the nice Customs Officer returned. He didn't say a damn thing. He took his seat in front of me, and picked up his 'franking stamp" and proceeded to stamp my passport, stamp my work permit, stamp my airline ticket. He then looked up at me and slid my docs and passport back to me underneath the glass and he the said: "go."

That was my friendly welcome to Moscow. Apparently, everything was just hunky dory, but I didn't buy it for one minute.

I figured there would be a tail on me wherever I went from that moment on. My work permit was officially authorized so that part was no problem. I was legit, and my Canadian Passport was a good thing to have. The part about me living in the USA, well I guess that too was okay, and the airport probably didn't have records of Hungarian revolutionary "Wanteds" on hand regarding my father.

I cleared customs as far as I knew. The strange part after all that was obtaining my luggage. I had one huge suitcase and a couple of smaller ones. Nobody wanted to see my luggage. Once I was done with Officer Russia at that Customs kiosk, there was no further checking of anything.

So I just wheeled my luggage cart out to the terminal lobby and looked for my hotel contact and pickup. Someone from the hotel was supposed to be waiting for me holding a sign with my name on it. Nope, that was not happening. I waited around for at least an hour and a half and nobody.

Well, this was just great. I didn't speak a word of Russian. Nobody that I came across seemed to speak English, and even if I wanted to make a phone call to the hotel, I couldn't because I didn't have any Russian currency coins called kopeks for their pay phones. Now I was temporarily stranded at the airport and feeling just a tad lost.

A few Russian mafia types approached me speaking Russian, probably wanting to offer me a ride, but I more-less blew 'em off. I wasn't about to chance being taken to who knows where. I walked the length of the airport terminal. It had some shops and some open counters all staffed, but I still couldn't find even one person who spoke English or was willing to engage me in English. But persevere I did, and finally found a counter with a nice lady who spoke English to whom I explained my situation.

She offered to call the Aerostar Hotel for me since I had no kopeks for the pay phone. She handed me the phone over the counter, and I was at long last able to speak to the Human Resources department at the hotel.

Well, it turns out that they thought I was arriving the next day. Everything was solved, and the hotel dispatched a hotel car and driver to get me.

A half hour later I was riding in the hotel car, down the highway from the airport leading into the city. My driver wasn't the greatest at English either but he did speak a few words. I started to feel a little, but just a little bit better. I was just glad as hell that I made it through Customs and not being held in some small room with electrodes clamped to my nipples and genitals. Okay, maybe that was stretching things, but still again!

I had begun my adventure, and now it was in full swing. I was in Moscow; it had finally hit me; I was in fucking Russia!

The hotel car driver sped me through the streets of Moscow heading down Lenningradsky Prospekt; one of the main arteries leading into the city.

My eyes were wide, and I took in everything that I could as we got closer into Moscow. Traffic here was just as congested as anywhere else I had been, but there was one major exception the scared the crap out of me.

It seemed to me that my driver pretended to be in some sort of a video game, where it was car versus pedestrian. He would no way slow down for anyone crossing the streets! I was freaking out a little, whereby I would brace myself expecting a sudden screeching skidding to a stop without seat belts to be found anywhere in the car.

The drive to the hotel was hair-raising, to say the least, he had absolutely no time or use for pedestrians, wouldn't slow down, and it became apparent very clearly that in Moscow it was the pedestrians that needed to avoid the cars, not the cars to avoid the pedestrians. If you've ever been to Moscow, you might know what I am talking about.

Driving into the city. However, it was evident that I was about to face a coming culture shock of sorts. Back in the USA, I lived in south Florida just a bit north of West Palm Beach, and I had just recently left the Virgin Islands.

Well, Moscow was a whole lot different than either of those two places. I initially found the city dirty from my first impressions, even just looking out the car's window, it wasn't a rosy picture. I also noticed that there were war memorials just about everywhere commemorating the World War II and the defense of Moscow from the Nazi invasion.

In 1941 the Germans came within 19 miles of The Kremlin, and it seemed to me that the people of Moscow were still reliving the second world war each and every day ever since December 4th, 1941.

Later, having already lived in Moscow for a few months I learned to appreciate just what they went through and from a purely historical and academic point of view I understood better as time passed.

One scene that stood out in my mind and is still in my memory was while driving to the hotel, looking out the window from my front passenger seat, we approached an intersection; where a group of people cued in line, holding onto burlap sacks. On the opposite corner, was a dump truck with its box lifted high having dumped a stack of potatoes into the street. I guess it must have been the dump truck driver, looked to me like he was selling the potatoes to everyone who stood in line, filling up their sacks and bags. Sort of an improvised instant farmer's market for potatoes right there and then and probably soon to be gone once the potatoes were all purchased. I didn't really make much of it, just seemed so different to me, but I was too consumed just taking everything else in as we whisked through the streets playing chicken with pedestrians. Jaywalkers beware!

At this point, I had no idea what to expect. So far I wasn't exactly impressed, but I certainly was consumed.

Other things that stood out were the enormous apartment blocks. It turns out that there are no single family homes to be found in Moscow.

I was there for seven months, and I never came across even one! On the outskirts of the city there are, but not in Moscow which has eleven million people! Everyone lives in apartment blocks. That in itself was truly unique to me.

So, on we drive, we are getting closer now to the hotel. I am now getting hungry and could most definitely use a bite to eat.

We arrive at the hotel, and the driver brings the car to the main entrance. I immediately am impressed. Greeting my car is a doorman dressed in topcoat with a top hat.

This looked like it might be a "class joint." In all seriousness, the inside of the hotel was most definitely high class.

My luggage was immediately taken care of by a bellman, and I was directed to the front desk of the hotel by another lobby attendant. Walking through the large double glass doors and entering the lobby was magnificent! Marble floors, marble pillars inside, the lobby, was expansive, and there were many guests milling about and ready for this? Everyone spoke English!

I made my way towards the front desk. It was quite a large area of the lobby with several stations for guests to check in as well as some cashier stations, serving guests who were checking out. There also was a concierge desk. I noticed immediately that all of the employees were sharply dressed in well-tailored uniforms and everyone was well groomed. Now why is all this important? It's very significant because this was not the norm in Moscow, not at all! The norm was a one-eighty from what I was looking at.

You see, in 1992 in the city of Moscow, with eleven million people, there were, count them, six, yes six hotels in the entire city. Okay, I need to qualify that. The six hotels I am referring to were: The Penta, The Radisson Slavanskya, The Metropole which happened to be quite famous and a landmark in the city. Then there was the Novotel, The Kempinski Baltschug and The Moscow Aerostar; to where I had just arrived. These six hotels were the only and I emphasize, the only properties that you would want to stay.

Yes, Moscow had huge humongous Russian style hotels; even hotels with three thousand rooms, but guess what, if you needed to use the washroom, well it was a not too far walk outside your room, down the corridor to the end of the hall and good luck finding toilet paper when you got there.

So, in essence, no Western business person from England or Germany or Canada or the USA or even Hungary would want to stay at one of the typical Russian hotels.

The Moscow Aerostar Hotel was a jewel in the crown so to speak of the newer hotels that were opening or slated to open as new capital was being pumped into the new Russian economy. All the six hotels in the city were pretty much full all of the time. With the Soviet Union being no more; investment was pouring into the city and hotels were being constructed faster than you could fill them.

I walked up to the front desk, and I couldn't get over how elegant and professional the front desk staff was.

The women were all gorgeous, and all the staff spoke English. So now here is a stunning fact for you…Russians were not allowed to rent rooms in the hotel or stay in the hotel. Isn't that just incredible? Russians could visit, but only to visit a guest, who had business to do with them, but a Russian off the street was not allowed in the hotel, not even to come in for a drink in the lounge.

This I found hard to believe. Here was a hotel in Moscow and a resident of the city could not come inside, unbelievable! That little tidbit of a fact I came to learn within a day or two of being there.

So I check into the hotel and am given a standard room where I was to live for the foreseeable future. The room was well appointed with furnishings, but it was on the small side.

Then again, I was only going to use it to sleep anyway and I had a whole city to explore and a way of life to learn. I wasn't too concerned about the size of my room.

I turned on the TV and much to my surprise, and pleasantly so, there was CNN with Larry King conducting an interview.

The Aerostar Hotel had CNN! I was floored! Hey, this is great!

In fact I couldn't believe it and not only was it CNN but it was live, like the real thing, not taped and then re-broadcasted, it was live from the USA.

Christ almighty, I almost felt like I was back home in Florida! Okay, maybe not, but you can imagine my surprise.

Next, my stomach started speaking to me again, and I remembered that I was starving. I hadn't had any food for a good while, so I found the room service menu and ordered, and yes, the room service order taker on the phone also spoke English.

Now things get a little interesting for my first few hours and my first day. About a half hour later a knock comes on my door, and it's the room service waiter with my toasted club on rye, a slice of black forest cake and coffee.

The room service waiter is a very nice gentleman who also speaks English, but broken English, but enough to get him by.

We engage in a little friendly conversation, and I introduce myself and tell him I will be in the hotel for a few months doing special work for the hotel accounting department. He welcomes me and then I ask him how long he'd been working at the hotel. He tells me just a few weeks, and of course I ask where he worked before and what he did. I almost fell over when he told me what he did before becoming a Room Service Waiter!

He was and is a Rocket Scientist for Christ's sake! Yes, you read that right.. a friggin rocket scientist!

Yup, he left his government job as a rocket scientist with the Russian space program and took this job in the hotel as a full-time Room Service Waiter because here in the hotel he could earn foreign currency in the form of tips. I tipped him two dollars U.S. cash. In one day he was able to earn at least fifty U.S. Dollars in tips. His regular hotel pay was in Russian rubles which amount to almost nothing.

You see, as a government employee; Rocket Scientist for the national space program his monthly take home pay was the equivalent of approximately twenty-five U.S.

Dollars. Did you get that? Twenty-five bucks a month, whereas working at the hotel, he earned at least fifty U.S. bucks a day being a room service waiter.

At that particular time, the situation was unclear as to whether Russian nationals could possess foreign currency or not. Many things were in flux and laws had not yet been enacted to cover such things as holding foreign currency. So, there you go. Living and breathing evidence of an internal national brain-drain that ran through every thread of Russian society with the collapse of communism and the arrival of capitalism.

So, my room service waiter was a rocket scientist. I wondered what professions the other employees held before becoming employees of The Aerostar. Turns out that my soon to be Russian girlfriend who was the sales secretary in the hotel, but before taking the sales position at the hotel, she was a Linguist Teacher at Moscow University teaching English and German. I spent the next seven months in Moscow; that was my first day. The welcome at the airport wasn't so great, but the welcome at the hotel was superb.

Welcome to Russia Frank.

***

MOSAIC LIFE TILES

Earthworms to Titleist

Oh jeez, where do I start? I guess I will start with us arriving in Canada in 1957. It was just the three of us. Me, I was six years old, my father was thirty-four years old, and my mother was twenty-seven. Well, to be fair there really were four of us, but my brother like I mentioned in a previous life tile, was still living inside my Mom, but ready to pop any day, so he was like, minus one-month-old. Although my baby brother was born in Canada, he was still Hungarian in my mind because he was made in Hungary, he too traveled and came through the same ordeal and terror as we all did, but of course, he was much more comfortable where he was.

We were what you would now call "political refugees." Well, maybe me not so much because of course, I had no idea what politics was all about, but I did know one thing though, and that was if you are on the wrong side of the fence, politics can get you killed.

We had escaped from Hungary during and because of the Hungarian Revolution of October 1956, and after waiting for almost a half a year in refugee camps and temporary housing in Austria, our small family was finally given the green light for immigration to Canada and a new life in the new world.

On the day we arrived in the spring of 1957, we were to become new Canadians. Neither my mother or father spoke any English, we all only spoke Hungarian. You know, now looking back on things, knowing that we were waiting in Austria for travel passes through the Red Cross to come to Canada and had been waiting for such a long time, the planning could have been a lot better.

The fact that Canada was willing to accept refugees from Hungary and the fact that Austria was a staging ground for a half year and more, there certainly wasn't much thought given to preparing for the transition.

The one thing and the most important factor in one's ability to blend and assimilate to a new culture was the language barrier. With all that time having been spent in Austria waiting for months on end for the go-ahead to Canada, the authorities could have made things a hell of a lot easier for everyone by offering conversational English classes. That alone would have been huge! But no, there we were in the holding camps, thousands upon thousands of Hungarian refugees, all waiting for tickets and a new life to be in western Europe, or across the ocean to the USA or Canada and nobody spoke English. And nor were they going to before leaving the camps. So we arrive via military aircraft in Montreal and speak not a word of English or French for that matter.

How things have changed since those times. Canada has since woken up to the plight of refugees and has long ago implemented a much more realistic and I must say compassionate assistance programs for newcomers.

Newcomers to Canada now have a multitude of government and community coordinated programs to help assimilate into Canadian society. When we arrived; not so much.

Actually, to be honest, almost nothing and if there were any such programs as language immersion or assisted temporary housing, we certainly were not privy to that or made aware of it.

So just put yourself in this position whereby you arrive on foreign soil, are placed into a hotel with your family for two weeks, yes, just two weeks and afterward are evicted and told to find your own source of accommodation without a job and not knowing the language.

This was the Canadian plan of action for newcomers to Canada, oh and of course we were now to become "Subjects of The Queen."

Although I hadn't ever given it much thought over the years, and I truly am proud now to be a Canadian Citizen, naturalized albeit, but proud to be Canadian. Now having said that, I do have an issue, personal belief and opinion, but an issue nevertheless. So, here it is, see what you think.

I do somehow find it to be a violation of human rights in general that a free person is required to swear allegiance to another human being, that in this case being The Queen.

I cannot square this one bit. If I am to be free, then I should not logically be bound to swear allegiance to anyone, except perhaps to my Maker and that too is an individual choice not to be challenged by any one person or government.

Recently in Canada, the term 'Subject" was replaced by the words: "Commonwealth Citizen," in 1977 by the Citizenship Act but all Naturalized Canadian Citizens are still required to pledge an oath of allegiance and fealty to the Monarch, that being the Queen or King.

In my mind, the Monarch is just another person, and that is where my comfort zone starts feeling uncomfortable.

Just thought I would throw that in there. I suppose I have an issue with the whole Monarchy thing. But come on now… you do have to admit; you too probably feel a little uncomfortable in your own skin knowing that you have an overlord you are obligated to; that being The Queen or King of England and in this case Queen of Canada if in fact, you happen to be a Canadian reading this book.

I have no problem with swearing an allegiance to my country, but to a person? Well, that to me is tantamount to bordering on shades of slavery. It's along the same lines as the traditional wedding vows that had the bride "obeying" her husband. That, of course, was changed after smarter heads prevailed on the matter. I think the swearing in of naturalized citizens in Canada should be changed to eliminate allegiance to a Monarch if it hasn't yet been and keep it allegiance to your country instead.

Okay, I got off track there a little but probably worth the side trip; love to inject a little provocation and a pause for inner reflection as a story is told.

I think it's good to stoke the dying embers of controversial issues and bring those to the surface glowing again in your own mind to see if you are comfortable with it, or has it taken you by surprise and now feeling uneasy as to why you haven't examined your stance on the issue; or it may not be an issue at all for you, is it?

The Canadian government tried to spread the newcomers across the country according to their professional and vocational backgrounds. Those who were farmers were placed on trains and sent to Manitoba, Saskatchewan, and Alberta. Those who had industrial backgrounds and worked in factories or had machine skills, mechanics etcetera were sent to Hamilton, Toronto and Sault St. Marie and those with academics, doctors, engineers, and teachers were dispersed throughout the country to the areas most in need of these skills.

My father happened to be a tradesman, a machinist. My mother was an office secretary, but without language skills, her plight would be relegated to common labor or domestic worker, and with her ready to give birth any day, she would be a housewife for the foreseeable future.

Being a wife was no problem, being a housewife was a totally different thing. We had no "house," and facing eviction from the temporary hotel accommodation, we didn't even have a cardboard box on the street to call home. My mother couldn't even be a "boxwife" let alone a housewife.

With my father being a machinist, the government had us further going on to Hamilton which was a major industry center in Canada where his chances of finding work was to be best according to the Canadian government at that time.

So for two weeks, we were placed in a hotel on King William Street, just down from where the City Centre Farmers' Market is located today. Back then it was mainly an open air market as I recall, but since has been modernized.

We were in the downtown location which did give my Dad access to the immigrant community which was evident in the city center. My Dad being a pretty outgoing and friendly sort of person managed to tap into the local immigrant community.

Having kept his eyes wide open and ears to the ground, he found himself being acquainted with a local Hungarian businessman who owned a sewing machine shop on the corner of Cannon and James Streets, just a couple of blocks from where our temporary hotel was located.

Without any government assistance and no money to be had we were at the mercy of charity, but that was not something my parents were comfortable with and weren't about to accept without some sort of arrangement for repayment. My parents were much too proud for that.

This sewing machine retail shop owner, his name was Martin; he in fact, did befriend our family, it turns out had what you might call an apartment. But not really, it was better described as "just space" above his shop and offered it to my father and mother as a place we could live until we were more established. So we grabbed it.

It wasn't furnished except for a rickety old dresser and mattress without a bed, I think it had some sort of a kitchen, it must have. We had no money of course but needed money. Martin explained things that helped us out over the course of the coming weeks, and he took us around the city in his car to get us familiarized with Hamilton.

I remember a funny incident one day while Martin was taking us on a car ride. At that point I still hadn't had too many car rides and every opportunity to have another one was huge!

This particular afternoon, Martin had decided to take us out for a car ride, and he was going to buy us something to eat. Well, this to me sounded like quite the adventure, a car ride and something to eat! It then turns out that he was to treat my Mom and Dad and me to "hot dogs." Well, we still, of course, had not learned any English, maybe my Mom and Dad already knew a few words at least enough to say hello and thank you and probably, "sorry don't speak English." So naturally we wouldn't have known the words; "hot dog."

Martin decided he would translate it to Hungarian and having done that I was shocked to learn that we were going to be eating hot dogs. Well hell, I wasn't really sure what to make of that!

I remember asking my Mom and Dad if we were really going to be eating dogs. To be honest, I don't think my parents knew what to say because they didn't know either; they just knew that Martin was going to treat us to eating hot dogs.

Well, we drove from downtown Hamilton out to Burlington, down by Hamilton Bay. The only part of that trip I do recall was that when we finally got our "hot dogs" my first bite was a taste sensation!

You see, I don't think either my Mom or my Dad had ever had "relish" before. Mustard I knew, but "relish and ketchup" well, that was something totally new, and I loved it! I don't think I had ever had wieners before either. So on that day, "hot dogs" became my favorite food of all time!

My brother Jim was born a few weeks after our arrival to Hamilton. We had no furniture, no crib for him to be placed into.

My mother had no "Huggies," or "Pampers," or even regular diapers for my brother. I honestly don't know what she did. I know she must have got some from someplace. I do know this, however, and this will explain just how destitute and poor we were; my new baby brother slept in an open suitcase.

I remember that like it was yesterday! Having come home from the hospital with a new baby son, my brother's crib was one-half of a suitcase.

Martin, although he provided us with a place to live, was the cheapest bastard in the world. Yup, he bought us hot dogs, but I can assure you he probably kept a ledger somewhere and made my Dad pay it all back.

He most definitely could have provided some sort of furnishings, or gotten us in touch with somebody who might have had a crib but no.

Over the years to come, I remember he had again taken advantage of my Mom and Dad. We moved from the space above his shop to a single family house he owned out in East Hamilton which also had no furniture.

By this time, however, my father had found himself some work. It was fortunate that we arrived in the springtime to Canada; otherwise, I have no idea how we would have coped.

Martin put my Dad in contact with a man who operated an "earthworm picking business."

Now I don't know if you have ever heard of "earthworm pickers," but believe me, it was a job that did exist.

Nowadays earthworms are still utilized as a product, but I believe the cultivation and production is now farmed.

Back in the 1950's, it wasn't like that. I guess there was a local businessman who had arrangements with golf courses in the region whereby he could take a team of people out onto the golf course at night and pick the earthworms that came to the surface.

Earthworms were sold by the business to bait shops and distributed nationally to pet shops as food for various other animals but mainly used as bait for fishing.

So that spring and summer my father would spend his nights being a worm picker on a number of different golf courses within the Hamilton region.

Once my mother had recovered after childbirth, she too would join my Dad each night, going to pick worms all night long.

You see, although my Dad was a certified machinist, he was having no luck in obtaining employment in his line because of the language barrier. He would need to learn English first and same with my mother. But it wasn't like things are today.

Today refugees arrive, they are provided housing and given time to learn English as well as spending money, and after that perhaps a year or so; they would be in a position to seek some sort of employment in their line of expertise.

But no, my Mom and Dad were worm pickers at night. You might ask, well who then took care of my baby brother Jim while your parents were out working all night. Answer: me!

I cannot even begin to imagine just how much it must have torn my mother and father apart to leave me looking after my brother who was just a few months old. But my parents had no choice. They needed money, and it was the only form of employment available to them at that time. It was an extremely difficult start for us in Canada. I am certain we suffered more than our fair share in our first couple of years.

My parents would leave just before midnight and not be back till five or six AM. I was taught how to feed my brother with a bottle and how to change his diaper if he was crying a lot. I was only six years old, and I suppose I had already crossed minefields, escaped tanks, and crawled through the barbed wire; giving my baby brother a bottle would not be that difficult for a war veteran like me!

Down the street on James Street, not more than maybe a half a block up from where the sewing machine shop was located there was a furniture store. Once my parents had some money, they were granted credit at that store down the street, and they purchased some bare necessities on an installment plan. There was one thing at the furniture store that I found to be incredibly fascinating.

During the daytime, since it was summer and I wasn't required to attend school; well I would spend most of my time standing in front of the furniture store window or sometimes I would go inside and stay as long as I could until the owner got tired of me and kicked me out. Can you imagine what I was so fascinated about?

Well, I was six years old, and I never had in my life before coming to Canada seen television!

Inside the store window the shop owner had a line of televisions and on one of the sets, there was sure to be cartoons!

Holy smokes, another incredible thing in this new country! Well, I was glued to the televisions in the window and sometimes I would sneak inside and sit on a chair and watch. I think the owner of the store must have taken a liking to me because he would allow me to stick around for quite a long time. Also, my parents had purchased some furniture from him, so I guess he tolerated me.

As time went on, we relocated to Martin's single-family house in East Hamilton, by which time my father had secured some other form of employment, with a construction company that was clearing land for new subdivisions up on Hamilton Mountain.

With the extra money, he was able to finally buy our own very simple TV set.

He also bought his first car that he ever owned in his life. It was a 1949 Ford as I recall. Well, things were looking up! My mother and father still moonlighted even at this later time picking worms, which they continued doing for another four or five years I believe to supplement my Dad's income.

My mother took various odd jobs cleaning houses, and I remember for a while she worked as a laborer in a canning factory, which she had to give up because it involved lifting boxes of canned tomatoes all damn day long.

My father worked the graveyard shift eventually at a transport company, working as a mechanic, maintaining the fleet of eighteen-wheelers for Hanson Transport. It was a large trucking company that later went belly up.

For me, well we had a television so my world was complete! I could watch cartoons, and I could watch cowboys and Indians.

I remember watching a show called "The Cisco Kid," and of course I was totally floored when I came across a movie called "The Mummy."

I was even more elated when I learned that turning the dial on the TV would bring about something different, a different show. I was intrigued how I could turn the dial back and forth and go back to the show I was just watching. This was pure magic!

I was just a kid, so I was speaking English before anyone else in my family. For a little guy like me, it was easy to pick up the language; kids are like that apparently.

Then I remember one day, our family took a car ride very first time, in our own car, just a few days after my parents bought the car and we drove from Hamilton along the two-lane Queen Elizabeth Highway to Niagara Falls!

This was another sight for me to behold. Well, I'm sure we all marveled at The Falls as everyone does who sees it for the first time.

For me, my life was complete. On that day I was wearing a cowboy hat with a whistle, I had an ice cream cone, I had a toy cap pistol with holster tied with string to my pant leg, I had seen the greatest wonders of the world; cartoons, Cowboys and Indians on TV, and Niagara Falls. I was now fast becoming a real Canadian!

The years passed by and both my brother and I grew up, and my mother and father eventually became voting Canadians as citizens. My brother remained living in Canada except for one little jaunt he took working in Bermuda, and I eventually immigrated to the United States through my marriage to an American. Later on, I did come back to Canada and stayed.

The years went on and on, and in the month of March of 1990, my father came to visit me on his own while I lived in Florida.

At that time I was working for a large hotel management company in their corporate offices as one of their executives. The hotel company was centered in Stuart Florida, and one of the properties it managed was a fabulous resort called The Indian River Plantation on Hutchinson Island located on the Intracoastal Waterway.

The resort was quite lavish and naturally had an 18 hole golf course.

I had lots of perks working for the company, and one of my perks was to play golf for free, including club rental, golf cart, and I could bring a guest, no problem.

Well after my mother and father had stopped picking worms on golf courses he never set foot on a golf course again, until he came to Florida to visit me! For the first time in his entire life, when my father bent down to touch the turf on a golf course, he no longer reached for a slimy worm, but a Titleist golf ball! He started crying, and so did I.

I hugged him and told him I loved him and said, "Dad let's play golf."

We did; we played golf that afternoon, but sadly two weeks later a heart attack took my Dad, and he passed away.

Thankfully, the good Lord at long last had allowed him to use a golf course in the way it was meant to be; playing with his son on a sunny day, swinging a three iron at a Titleist.

***

A Perfect Match

I had absolutely no idea that the events of this day were to lead me to an epiphany. I suppose that is how epiphanies come about; unexpectedly. I was nineteen years of age, and it was winter. I still lived at home with my parents. We lived out in the country, in the Niagara Peninsula.

We had a small fruit farm. My Dad worked in the city; my mother had a part time job in the housekeeping department at the hospital in Grimsby a half hour drive from our small hobby farm. I didn't have a job at all, and I was for all intents and purposes done with high school.

I was facing a problem. The problem was; my future. My parents didn't have the resources for me to continue with my further schooling, such as community college or even for me to enroll in a trade school and become a tradesman. My father had a plan for me, which was for me to work in the same machine shop where he worked. He said he could probably get me in as a machine apprentice.

Well, that had no appeal to me whatsoever. I wasn't a mechanically inclined sort of person. On the other hand, I didn't know what I wanted to do for the next few months or the coming year and more importantly for the rest of my life. I was going through a period of frustration.

I looked at the friends I had and what they were doing with their lives. Most of the kids from high school class were going on to further education. My close friends, it turned out were finding jobs in the city, being Hamilton. One of my friends found himself a job as an air conditioning apprentice. Another friend of mine was working for his father in his janitorial business, and another was pumping gas at a Canadian Tire gas bar.

I didn't have friends or peers, who were going on to University, and I wasn't going to medical school nor was I cut out to be an engineer. Interestingly enough though, I had a certain level of intellectual curiosity, and I felt I could do something interesting and challenging, I just had to find the right vehicle.

There was something burning inside of me that demanded some sort of direction. That burning desire grew hotter and hotter as the winter went by, and I needed to find my direction, and a clear path to my future and I needed to find it soon, I was becoming very restless.

I knew that if I didn't find my path soon, I would have no choice but to go work with my Dad at the machine factory in Dundas and that I didn't want to do at any cost. My candle was burning at both ends. Time was running out; soon my father would insist that I either go and work at the factory with him or I find myself a job somewhere doing something because I was going to be twenty years of age and I needed to become a man.

I had done odd jobs throughout the year, working part time on neighboring farms, either working with tractors, tilling soil between rows of grape vines or picking fruit, but those were all fair weather jobs.

During the winter there was no part-time work to be had. I did have a few bucks tucked away from the work I had done during the summertime, and I did have a car, not a good one, but I was able to get around.

I had some friends I hung around with as well. It was the winter time. I needed to get out now and then and spread my wings, just being with my friends.

My two close friends were Mike, whom I had known from the age of six, and Tim, whom I had met just a couple of years ago through Mike. We seemed to mix well, and we would hang out, play pool, go to the movies, and try reducing boredom.

This particular night we decided to take a drive into downtown Hamilton and discussing what we would do, we decided to take in a movie. The movie was at the Odeon theater. Beside the Odeon; was an indoor parking garage attached to The Holiday Inn Hotel.

This was the night my life changed, at least it did in my head, and soon after that, the epiphany started taking shape.

We arrived in Hamilton and parked at The Holiday Inn. Now this was significant. The only thing that interested me at The Holiday Inn was the parking garage because it was conveniently located and parking was free after 6:00 PM.

Growing up; our family never had a need or use for a hotel. We never even went out to a restaurant for that matter and staying at a hotel was just not ever in the cards. Our family did take vacations, but we were campers. In the summertime, we spent a couple of weeks camping in the Kawarthas on Jacks Lake. We could afford a tent and campgrounds, but not a hotel. We weren't the hop on a plane and go to a Mexican or Florida resort types. We had never stayed in a hotel, except for the time we immigrated to

Canada and were put up in a hotel temporarily as a place to live for a few weeks. Other than that I hadn't even ever set my foot inside a hotel, and I was already almost twenty years old.

We parked the car, but the movie wasn't to start for almost another hour. The three of us had time to kill. We decided to go for a drink, since we had never been to the bar at The Holiday Inn. Well, it was just right there, so why not check it out; see what it's like.

We walked in through the front doors of the Holiday Inn. My two buddies went to the washroom, and I was left alone standing around in the lobby. I wasn't thinking about anything, in particular, just waiting for my buddies to come back, but I was looking around being observant and something caught my eye. I noticed that down the hallway off the lobby, leading to the restaurant there was a counter with a brochure rack.

I walked up to the rack; the rack contained individual separate slots with different brochures in each slot. It turns out that each brochure was about a different Holiday Inn located in towns, cities, and communities throughout Ontario and one section of the rack that contained an entire booklet. This booklet was a worldwide directory to all the Holiday Inns Internationally.

There was also another row of slots that contained colorful glossy brochures of Holiday Inns throughout the Caribbean; resort properties. I took a few of those brochures along with a worldwide directory booklet, still not very interested but they called out to me for some reason, and I now know what that reason was, but it hadn't hit me back then.

I mentioned that our family went camping in the summertime. We had started doing that when I was still little. In order to go camping our route would take us along The Queen Elizabeth Way highway, then the 427 north that would then connect to the 401 East through Toronto onto Oshawa and then Highway #115 into Peterborough and eventually to our campsite in the Kawarthas.

The reason I bring this up is because every time we took that trip along those highways in the sixties, there was one thing that I always looked forward to seeing and that was the Holiday Inn marquis.

Back in the sixties, The Holiday Inn Hotels had these enormous signs with flashing lights around the perimeter scrolling and pointing to the hotel entrance.

The signs were concrete and quite large, visible from the highway, no problem, and as a kid, I remember I looked forward to seeing these signs along our trip.

I knew where each Holiday Inn was located along our trip.

The first one was on Highway 427; later I found out the Holiday Inn was actually called "Holiday Inn 427." Then next one just past the Rexdale shopping center that you could see from the 401 and another farther along the 401 towards Markham, and there was another in Oshawa off the highway, and the last one in Peterborough.

These Holiday Inn Hotels were etched in my mind even as a kid, and it was all because of their signs.

Also, I liked the name "Holiday Inn." It made me think about having a holiday on the inside of a building. It's funny how the mind works sometimes, but that's how I thought of Holiday Inns as a kid, never having set foot inside of one until that day came about when I was collecting the Holiday Inn brochures from the brochure rack at the Hamilton Holiday Inn.

My buddies were still in the washroom I guess, because I went and sat down in the lobby onto a nice plush leather sofa.

I remember this now as if it was yesterday. I was sitting on the sofa and a couple, man, and woman came out of the elevator having come down from the indoor parking garage, carrying their luggage and walked up to the front desk.

Well, I suppose he was always around somewhere, but I hadn't noticed him until this couple walked up to the front desk, but it was a bellman. He had a nice red uniform and took a position just slightly off to the side of the couple at the front desk and greeted the man and woman.

I then watched, as a young man, not much older than I was, maybe by a year or two, who was behind the front desk, he was the Front Desk Clerk, he also greeted and welcomed the couple to the hotel.

The guests went through the checking in procedure, and the bellman took their luggage, placed it onto the luggage cart and escorted the couple to their room which was to be accessed using the elevators.

Well, I sat on the sofa, I was taking in all of the proceedings. The young man on the front desk appeared to be very efficient; he was running the show! He was in charge of everything!

I looked at the brochures I had in my hand. I must have had fifteen brochures, of all the different Holiday Inns throughout Ontario, and then to top things off, I had brochures with Holiday Inns all over the world, and looked at those brochures of the beautiful Holiday Inns in the Caribbean, in Bermuda, in Jamaica, in Grenada and The Bahamas! Wow, Holiday Inns were all over the world, and my world directory contained hundreds and hundreds of Holiday Inns!

I was a "map traveler." As a teenager, I would devour Atlas after Atlas. I knew where most countries of the world were geographically. Their capital cities, rivers, lakes, mountain ranges and more. Geography was my favorite of all subjects, and these hotels were located in some of my most favorite places.

Sitting on that sofa I was beginning to "feel it," I remember so well.

I watched as the front desk clerk was going about doing his job. He was probably checking reservations to come while answering the phone and dealing with a number of things; and he was dressed so sharp!

I watched him, and I said to myself, "I could do that job no problem!"

Then after having said that to myself, something incredible happened, and to this day I am convinced it was meant to be!

Sometimes in life, you hit the lottery without hitting the lottery itself, but something happens to you that was meant to be; fate, destiny, an epiphany, call it what you will, but it happened to me.

Back in 1970, I had just taken up smoking; I think just about the entire world smoked back then. I resisted till I was about nineteen but I too got caught up in it. Anyway, I was pretty excited just sitting there on the sofa thinking about all this Holiday Inn stuff, and I decided to have a cigarette. I didn't have any matches with me to light my cigarette.

Well back in those days, you will recall that just about everywhere you went you could find complimentary little folding packets of matches lying around. In addition to that, every coffee table or side table had ashtrays. Smoking was big back then. I looked around for a pack of those flip open matches, and sure enough, there were a few on the side table beside my sofa.

I picked one up and lit my cigarette and watched the front desk clerk carry on with his duties all along thinking, how was I going to go about doing what he was doing? The answer came like a lightning bolt!

I was fiddling with the pack of matches in my hand when I looked at the pack and on the back of the pack was an advertisement. It read; Start your Career in the Hospitality Business: Lewis School of Hotel Management. Call for free consultation. And it had a toll free number to call in Toronto.

I couldn't believe what I was reading on this package of matches! All of a sudden I became so excited I didn't know what to do.

I started shaking. I knew instantly, that I had found my path!

My entire life had taken a sudden turn. I knew I wouldn't have to go work in a factory. I knew I wanted to work in a Holiday Inn in the Caribbean. I knew I wanted to be that guy on the front desk in front of me. I all of a sudden knew that there was a reason those Holiday Inn Marquis stood out in my mind on all those trips along the 427 and 401.

I looked for more matches with the ad, but the one in my hand was the one and only pack of matches with an ad for The Lewis School of Hotel Management.

All the other matches were just regular Holiday Inn freebie matches. This one lone pack of matches had been left there apparently by someone else; already half the matches were pulled from it anyway. I held onto that pack with my life that night.

My buddies and I went to the show that night, but my mind was on Holiday Inns all night long, and I could hardly wait until the following day to call the number on the match.

The next day I called the number and was told that they could arrange for a representative to come out to our house and give a presentation about the school and how things work. During that conversation, I was told that the hotel management course was available on campus, being onsite at the actual school in Toronto and could be completed in one year or I could choose an option of two years and do the course through correspondence which would cost less.

I asked how much the correspondence course was. I was told it was three hundred and fifty dollars!

Well, that was a hell of a lot of money back in 1970, but it could be paid by making monthly installments. I figured if I could take the course and get a job in a hotel, I could maybe pay for it.

After that phone call, I approached my parents about it all, and they agreed to have the representative come from the school and do the presentation.

My parents were good with everything, and they signed me up for the correspondence course. Everything had taken place within two weeks of my coming across the package of matches. My parents put a down payment on the course which allowed me to enroll. I received all of my course materials a few days later and started in, diligently!

The course came with textbooks and accompanying phonograph 33 rpm records. The first year were twelve records, one for each month covering different topics and aspects of the hotel business. Actually it was a lot of fun, and I immersed myself into the course. I submitted my lessons and was doing very well; my marks were all high.

After starting the course, within a few days, I went back to The Holiday Inn in Hamilton and applied for a job.

I don't know how, but I managed to fill out an application which was available in the offices of the Assistant Innkeeper and Catering Manager. As I was filling out my application, the Assistant Innkeeper walks into the office and takes a seat at his desk. Once I finished the paperwork he tells me he may as well look at it since I was already there sitting in front of him. He reads that I am currently enrolled in a hotel management course. He keys in on that and asks me about it. I tell him about the correspondence course and that my desire is to work in the hotel business because I am service minded and would very much enjoy working with the public.

He hired me that day as a "banquet porter."

I had my first job in the hotel business, and I was to start right there and then if I could stay.

I did. I started working right after I filled out my application! Back in those times in the early seventies, there was no HR department, it would have been called "Personnel" if one did exist, but in a mid-sized property, the Assistant Innkeeper or department heads did their own hiring of the staff they needed.

I was excited. It turned out that I didn't like being a banquet porter, but I stuck with it, never complained and worked hard, did everything I was asked to do.

The hotel General Manager; "Innkeeper," placed me onto a job rotation schedule. He had picked three young employees, with me being one, to be trained in various areas of the hotel, shipping and receiving, guest services bellman, and front desk clerk.

I trained in all areas successfully, and eight months after I started working at the Holiday Inn Hamilton, as a banquet porter, I had become a front desk clerk.

I was now wearing a suit; I had a sharp looking tie, and after a few weeks, I was "that guy," running the show on my own.

On the first day, that I was to be working the afternoon shift as a front desk clerk on my very own at the Holiday Inn, I had made sure to bring the pack of matches I found eight months ago with me to work.

Later that evening, when it was quiet, I came out from behind the front desk and took a seat in the lobby on that sofa I had been sitting on eight months earlier.

I took the pack of matches from my pocket and looking at the pack, placed the matches back onto the side table where I had found them and went back to work as a front desk clerk at the Holiday Inn of Hamilton Ontario. I was on my way to becoming an "Hotelier," it was a perfect *match*.

***

An Extra Shiny Tile

On writing this first series of "Life Tiles," along the planning and decision-making processes that went and even now go through my mind giving thought to what my next Life Tile will be about, certain events keep injection themselves into my train of thought.

No matter how hard I try to focus on a section or group of tiles that would make an interesting next short story; these other extra shiny tiles keep popping in and out of my mind yelling out, "write about me, write about me!" So I will and in doing so, I am hoping they will then be out once and for all and no longer get in my way, at least for this one tile I am about to write.

These extra shiny tiles probably don't require their own short stories.

But if you were to look at a mosaic of my life's events and stand back at a distance, you would no doubt see a number of tiles depicting my life that have a very different brilliance to them; being extra shiny almost with a light of their own.

Scattered around here and there, not really pulling anything together, but just shining very much on their own as part of a short story already told or waiting to be told, but needing its own special mention.

In no particular order, I will write about one now and then leave some for later in subsequent books to come.

The first one I find butting itself in line has to do with my time in Russia. It is not first because it deserves to be first, but I have to start somewhere. I cannot start with all of them at once. To be honest, I almost feel guilty of not being just, by having to start with one over the other because these "tiles" are so outstanding by themselves that I cannot put them into a sequence, at least not linear. So to put you in the picture, if you have already read the preceding short story, Welcome to Russia, then you would have a fair idea and insight to this "tile," but let me explain.

I found myself working in Russia and may as well say, living there. I could probably get away not claiming to have actually lived there if I had been staying in a hotel for the entire time, but that was not the case. I was allowed to live in the hotel where I worked, but after a month I was required to move out of the hotel and rent an apartment in the city.

I did that. I relocated from my hotel room to a typical Russian style apartment in a huge apartment block, and I lived there for six months having regular Russian neighbors down the hall and throughout the building. I believe I was the only foreigner in the entire apartment block.

So living in Russia and knowing I was going to be there for a good while, I settled in and tried my best to understand Russian society and looked forward to getting out and about in the community.

Each morning I would walk to the subway station which was a good fifteen-minute walk, and I would take the subway to work along with millions of other Muscovites and repeat the process coming back home after work. So, I, in essence, lived in Moscow, didn't just work there.

As the days and weeks went by I did my best to take in the city, but in reality, it just wasn't working as I wanted it to and the main reason was that I didn't speak Russian.

Once I left the hotel where I worked with English speaking employees, well out in Moscow virtually no one spoke English, and my Russian was not conversational at all. What I needed was a Russian friend who also spoke English.

Well, where was the best place to find such a person? Naturally, it was my workplace. Right about that time, when I was thinking along those lines, a natural human condition kicked in, and I was very desirous of having a girlfriend. So this now was making much sense to me; I'm in Russia and will be here for a good while, I live in the city and have my own apartment, might be a good idea to have a Russian girlfriend.

I was forty-one years of age, been divorced for three and a half years, had no kids, was single again and well, how should I say, I was "looking."

On top of that, I was rich!

Okay before you start rolling your eyes, let me explain that too.

Back in 1992 when Russia opened its borders to foreign investment with the creation of thousands of joint ventures almost overnight, the average income for the average Russian when converted to U.S. dollars was approximately twenty-five bucks per month! That deserves a repeat; per month!

I worked at the Moscow Aerostar Hotel as a Western hotel consultant, and although I wasn't earning a great deal of money by western standards, then again, mind you my apartment, meals and groceries were all picked up by the hotel, which was a major part of my contract. I was still earning thousands of dollars per month on top of that. Deposits in U.S. dollars were made bi-weekly into my Florida bank account by wire transfer.

Putting that into perspective now clears my suggestion that I was rich by Russian standards. In essence, I was earning in one day, what most Russians earned in say six or seven months!

I know that sounds unbelievable, but that was the case. In one day, I earned more than several Russians earned in a month.

Remember, the average income was approximately twenty-five dollars per month, when converted to US Dollars.

Keeping that in mind, there wasn't anything I couldn't do, or any place I couldn't go, because it might cost too much! There wasn't anything in the city that I couldn't afford to attend or do.

For example, the world famous Bolshoi Ballet was something I wanted to see and attend while I was in Russia.

Well, a ticket for a balcony seat cost an amazing twenty-five U.S. dollars. That was not a problem for me; I could go each and every day if I so chose. For a Russian to attend, well it was virtually out of the question. For the average Muscovite to purchase a regular ticket for the Bolshoi; that would be the equivalent of you having to spend your entire month's wages to see a ballet; absolutely impossible.

So how did the average Russian attend the Bolshoi you might ask? The theater was sold out every night.

It wasn't as if Russians weren't attending; they were; in droves, but for ninety percent of them, it was free.

When it was your factory's turn to be recognized for outstanding work and contribution to the Soviet system, then all the employees from that factory or plant or region or district were provided with free passes to the Bolshoi, and that only came about a few times in a lifetime!

Just because communism had come to an end officially, things pretty much for the average Russian went on as it did before, and most of Russia still worked in State-owned factories, and procedures did not change overnight.

A certain number of theater seats at the Bolshoi were reserved and held for foreigners and paying customers but most attendees, Russians were there as guests of the government and appreciation night for their hard work. They had a number of days to consume the government's generosity;

then it would be a different set of employees' turn from a different factory or region.

I on the other hand as I said, could attend anytime I wanted to; it was only twenty-five bucks!

Well after having worked at the hotel for a month or so, I had noticed from time to time a very attractive woman whom I guessed was in her late twenties. Hang on now; the hotel employed some extremely attractive and well spoken Russian women.

You see, to work at this particular hotel in a guest contact area, the Russian employees had to have some accomplished command of English.

This one particular young lady I learned worked in the hotel sales office, as a sales secretary.

From time to time, I had dealings with this gal because she would be the one to follow up on billing procedures and contractual agreements for groups and meetings held at the hotel and I would be in charge of actually collecting the balance owed.

My boss, at the hotel, who also was Canadian, but spoke Russian, was dating a Russian gal who also worked at the hotel.

I didn't waste any time figuring it would be okay to ask an employee out since I wasn't really an employee of the hotel, I was a consultant; so the rule about dating your co-worker didn't really apply to me. I had some flexibility in this area.

Well, I asked around a little, mainly my boss who knew everyone in the hotel and with him having a Russian girlfriend who also worked there, well it wasn't long before I learned that "Irina" whom I had an eye for, was not involved with anyone.

Now here was a bonus, Irina spoke fluent English, as a matter of fact, fluent German as well. She had been a Linguistic Teacher at the University of Moscow before taking the job at the hotel in the sales department.

The other thing about it all was that I found her to be extremely attractive and she had a very bubbly personality but make no mistake about it, strictly business.

In fact, I found her bubbly but abrupt. As soon as the topic had been addressed, her bubbly personality also disappeared, and she was off! The subject had been satisfied, and there was no need for small talk. She was all business.

I found that also to be appealing, not really sure why maybe it showed a level of confidence having no need to endear herself to anyone with niceties not pertaining to the subject at hand.

Well, how was I going to approach her without seeming to be interested in just one thing, and I wasn't interested in just one thing, I really and truly had a genuine desire in getting to know her. I liked her a lot!

One thing in my mind that kept creeping in was our age differential. I was forty-two, and she was twenty-six apparently, a good gap of sixteen years. Oh what the heck, not asking would always be a no, so I took the leap.

I waited to call her just before the end of the working day. There was this huge flea market in Moscow called the Izmailova flea market, it was a major attraction, and you could buy just about anything and everything you could almost imagine there, but you had to be careful not to get ripped off. Plus I had no idea how to get there, which subway trains to take and I really didn't need to get lost in Moscow.

At Izmailova, you could buy traditional icons there, war medals, Russian Matryoshka dolls, all kinds of knick knacks and some incredible finds if you looked hard enough. I wanted to visit this flea market and needed a guide, someone I could trust and someone with whom I could have a nice afternoon. This was my plan and here is how it went.

It was about a quarter to five in the afternoon when I called Irina in the hotel sales office.

I dialed her extension, she answered. "Good afternoon, sales office, Irina speaking, how may I help you?"

I responded. "Hi Irina, this is Frank in accounting, how are you this afternoon?"

She then said, "I am fine, thank you Frank, what can I do for you?"

I was now committed, so I went on. "Irina, I was wondering if you might accompany me to Izmailova Flea Market tomorrow afternoon, I think I would enjoy your company very much, and I could certainly use some help in getting around and maybe you could help me in finding good buys."

I had said my piece and waited for her response. There was silence.

Okay, I wasn't a kid any longer, and I had already been married once, and I had dated a few women in my time, even and had fallen in love a time or two.

But there, at that moment, waiting on the phone and hearing silence, I heard my heart beating.

The line was silent for a few more seconds and then I heard her voice, and she said, "sure why not, what time were you thinking?"

Once again, in her style, sort of very "matter of factly," but her answer was yes!

Wow, it was a lot easier than I thought it would be. Sometimes all you have to do is ask. It doesn't matter where in the world you might be, if you don't ask, the answer is always no.

Now things were easier. She wanted to know what time I was thinking of, so I asked if 1 PM would be good.

She agreed to meet me at the Metro Danamo Subway Station which was just across from the hotel, right at 1 PM.

I was elated! I would be meeting Irina and spending my afternoon with her.

The following morning I could hardly wait till the time came. It seemed like every passing minute was taking an hour. The time finally came, and I went to the subway station entrance, and there she was; waiting for me and smiling!

We enjoyed our day together, and I thought that it went well enough for me to ask her to join me for another outing soon. I later learned that it took Irina over an hour and a half just to come into Moscow every day.

First by local bus service from the town she lived in on the outskirts of Moscow and then by subway to the hotel where we both worked.

Irina and I became an item over the next couple of months.

Irina became my girlfriend, and to this very day, she lives in a very special place in my heart.

Whenever I think of her for more than a few moments, my entire experience living in Russia floods my memories within seconds, and I become overwhelmed with emotion.

Even now as I write this, I am doing my best to hold back my tears and not to break down. It has been over twenty years but my memory; it serves me as a time machine, and I am now there.

I was there in Russia during the winter time. One Saturday night, Irina and I decided to go for dinner at one of the better restaurants in Moscow, it was called Trenmoss.

We arrived late, but that was fine, the restaurant remained open late into the night.

After dinner, it was already almost midnight. Red square was not too far away, so we flagged down a car, yes, just a regular car, it was the Uber of the day.

Back then in Moscow you could flag down any civilian car driving on the roads and if the driver fancied stopping to pick you up, he would, and he would be your taxi ride.

Yes, there were regular taxis as well, but the public now took up an entrepreneurial spirit, and everyone got into the private taxi game to make a buck.

So we caught a ride to Red Square. Snow was falling gently, very gently but large flakes. We walked the cobblestones, past the huge rust-colored building that was the Lenin Museum on Red Square which more-less formed part of the street leading into the huge square. We walked through the snow, quiet as the night was, we walked arm in arm.

There was something that happened every hour in red square, and that was the hourly changing of the guard at Lenin's Tomb. It was a night that was meant for just Irina and me apparently.

We were the only two souls at midnight standing in front of Lenin's Tomb with the snow falling softly as we waited and witnessed the powerful and moving "changing of the guard."

As I stood there holding Irina, I stood beside her close. It wasn't cold that night, but I needed her close to me. I watched as the three guards marched slowly out from behind the Kremlin walls and marched in unison, and precision, carrying the entire power, history, and strength of the Soviet Union in every step they took as they approached Lenin's Tomb and changed positions with the guards already there.

As I watched, my mind raced, and my shining life tile now glows brighter than any other life tile of my fragmented life.

It glows alone, and its light will never fade. I had come full circle. I escaped Hungary with my mother and father. We escaped because Russia had imposed communism upon the Hungarian people. The uprising that resulted in the Hungarian Revolution of 1956 to oust our Russian overlords had my father blowing up Russian tanks, and killing Russian soldiers.

The revolution was put down by the Russian forces and Hungary once again brought to its knees. We ran for our lives and found freedom in Canada.

I watched as Changing of the Guards at Lenin's Tomb now was happening in front of me, and I held on to my wonderful and the most loving woman I had ever met.

She was Russian; I was Hungarian, and history was alive and crying in my mind as the snow fell and my tears would not stop flowing down my face. I had fallen deeply in love with Irina; she knew and understood totally what I was feeling at that moment.

Off in the distance, coming from St. Basil's Cathedral, in the quiet midnight air, someone started playing saxophone, and the sound of the brass instrument filled red square with a haunting and mesmerizing melody that I will never forget.

Governments can be so mean and warlike with opposing political views. Nations can be at odds with one another, but two humans in love can overcome all obstacles no matter what horrors history might hold. Love conquers all they say, and on that night I had conquered my fear of Russia. My Red Square Midnight Experience "life tile" burns brightly to this very day.

***

Mosaic Life Squares

My second book in this anthology series.
Some of the mosaic life squares I will be sharing and
a taste of things to come.

Walking with friends at 2:00 AM in the high Arctic on Baffin Island with the midnight sun brightly shining over ice flows formed on the Arctic Ocean

How driving across Canada in a TR7, on the Trans Canada Highway, with my brother; Jim, we almost bit it with a logging truck in a head-on collision. Still to this day, I'm not sure if my brother is even aware of this.

The time I was flying from Hawaii to Rarotonga; The Cook Islands, and the harrowing flight experienced over the Pacific.

Standing in Red Square in Moscow, wondering how in hell I ever ended up there.

Being chased by hungry wild dogs in the Arctic, only to reach the doors of a tunnel in the nick of time, before the dogs caught me.

The morning I was driving to work in my Corvair and suddenly having it catch on fire.

Taking off, on Northwest Airlines, only to have the takeoff aborted, twice, just before liftoff!

I was five years old, holding onto my Dad's hand, as we were sneaking and escaping through the forests of Hungary, escaping from the Russians, running for our lives.

Having to go to summer school.

Feeling bad about leaving my Dad, Mom and brother behind, when I left home and took a job in sunny Bermuda.

Sharing an apartment in Toronto with my brother Jim, the ups and downs of two brothers living together, dating a variety of women and never having enough money to do it right.

How I had to call hotel security to have a woman extracted from my room because she didn't want to leave me.

My first job in Bermuda and how I got it, "collect call" for a job inquiry.

The time I got stuck in a snow blizzard with my mother on our way to Florida on Interstate 79 outside of Erie Pennsylvania.

How I got myself into a predicament of having to tell my girlfriend that I had another girlfriend flying in for the afternoon; and how she bought it.

Snorkeling off Rose Island in the Bahamas, and thinking it was about as good as it can get.

Getting married in Las Vegas at the Candlelight Wedding Chapel.

Arriving in Caracas Venezuela thinking, I had entered a war zone.

Leaving Banff Alberta for a job in Fort St. John British Columbia, and then coming back to Banff a month later.

Getting stuck in Rarotonga, South Pacific, my passport was taken and no way out for two weeks.

Going to the first MacDonald's opened in the Soviet Union, and watching a half mile line of people cued to get served.

The time my brother drove into town to the video rental store, having only his t-shirt and underwear on, this was a funny incident.

How I met my wife the very first time I saw her and knew that I would marry her and how it came true.

Being caught on my houseboat in Hurricane Hole Marina, during a hurricane on Paradise Island in the Bahamas.

The night my brother and I were star gazing from my brother's boat, anchored at Beaujolais Island in Georgian Bay and settling on the fact that we were looking at the planet Jupiter, only to discover as dawn was breaking that it was just the mast light of an anchored sailboat further down the bay.

The time I decided to quit my job, on the spot (on an airplane) strictly on the basis of principle.

A moving moment in history for me, standing atop Gellert Hill, in Budapest, looking down from the Citadela. A cold February night, overlooking the brilliance of the Chain Bridge and the lights of Budapest, while holding on to my girlfriend Irina.

Climbing Mt. Ballyhoo in Dutch Harbor Alaska and looking out over the Bearing Sea, with visibility of 100 miles over the Aleutian Chain.

A word about job interviews, touching on some of the ones I've had, and lessons learned, some of the following but not all.

My job interview trip from Miami to Bonaire.
My job interview trip from Toronto to Chicago.
My job interview trip from Miami to Nantucket
My job interview trip from Miami to Dallas
My second job interview trip for a different job, Miami to Dallas
My job interview trip from West Palm Beach to San Francisco
My job interview trip from Banff to Toronto

My job interview from Toronto to Jim and Tammy Baker's Heritage USA resort.

My job interview trip from Toronto to Atlanta

My job interview trip from Toronto to Denver

My job interview trip from Toronto to Chicago

My job interview trip from West Palm Beach to Grand Rapids Michigan

My job interview trip from West Palm Beach to Mackinac Island

My job interview trip form Orland to Stuart Florida

My job interview trip from Miami to Orlando

My job interview trip from Banff to Fort St. John

My job interview trip from West Palm Beach to Caneel Bay US Virgin Islands

My job interview trip from West Palm Beach to Freeport Bahamas

My job interview trip from Miami to St. Maarten in the Caribbean

My job interview from West Palm Beach to Tampa Fl for Montana

My job interview from West Palm Beach to Park City Utah.

How I marveled at being able to finally see Budapest for myself, and how ever since 1993, I've wanted to return.

On being Hungarian.
On being a Canadian Hungarian.

On living in the USA as a Canadian.

Visiting a Bavarian German town in the middle of the rain forest in Venezuela of all places.

How I influenced some peoples' lives and made a difference.

Finding myself getting back into motorcycle and motorcycle touring.

Looking for a house on my return to Canada and settling on living in Northern Ontario.

Getting my mother's apartment rid of bed bugs (what an ordeal)

Living on Lake Nipissing in Northern Ontario and fishing every day.

Going to work in Turks and Caicos in the Caribbean from Ontario every month.

How I became a consultant for three years, in the Transportation sector, for the largest distributor of fresh cut flowers in the USA.

Having to leave my job in Puerto Rico and tend to my hurricane damaged house back in Florida. (Twice!)

My erotic and exotic Magen's Bay experience in St. Thomas US Virgin Islands.

How I learned my lesson; never to hire a friend to work for or with you and never to take a job working for a friend.

Going from Moscow Russia to Dutch Harbor Alaska in the Aleutian Chain.

How I became a professional hospitality consultant, held positions as Financial Controller and Internal Auditor, all of this without having finished high school.

My first job in the USA, on Miami Beach as Assistant Financial Controller, working for an owner who was one of the worst human beings I ever came across.

Cycling in 100 degree heat, in Florida with my brother.

My brother Jim and I enjoying the beach and Tiki bar on Singer Island in Florida.

On being a hotel management consultant for a failing hotel in Timmins Ontario.

Buying my park model trailer on Buckhorn Lake and living in it for five consecutive summers.

The adventures of my motorcycle trip in 2015 to the Gaspe in Quebec from Curve Lake Ontario, with two of my friends (and one other person).

On being an "immigrant family" and making a go of it in the new world of the 50's and 60's and on growing up as the years went by; living as new Canadians.

How I always figured that my brother Jim, had one up on me since he was born in Canada, therefore a "real" Canadian.

The ins and outs of arranging for a long term care facility for my mother and how the elderly are the most incredible people you will ever find.

How I sold an internet domain name that could have made me a millionaire but didn't.

How I sold another internet domain name that could have made me a millionaire again, but sold to the wrong company.

How moving back from the USA to Canada was a great decision.

****

Thank you for reading, and now I wish to say that this first book of my life tile anthology series now comes to an end. I hope the edges of my tiles found a way to touch the edges of yours or perhaps allowed for a momentary interchange of our lives' mosaic compositions.

***

The End

Dedicated
in loving memory

to

Izabella and Frank
my
Mom and Dad

I love you both forever.

Upcoming works in the works

MARCH 2017

MOSAIC LIFE SQUARES
MOVING AND PROVOCATIVE – ANTHOLOGY: 2

Picks up where Mosaic Life Tiles ended.

Frank Julius follows up with volume 2 of life's experiences.

Mosaic Life Squares takes the reader along with travels and adventures in the hospitality industry and other interesting, moving, juicy and colorful squares.

***

FRANK JULIUS
Books

BLOOD DICE

A crime thriller focused on the resort hotel and gaming industry set in The Caribbean.

A fast paced riveting Caribbean adventure, exposing corrupt governments and labor unions conspiring to take control of Holiday Jewel Hotels and Casinos' resort empire through murder and kidnapping. Follow Eldon Davis as he embarks on a forced adventure to rescue his daughter, save his company and bring the whole corrupt house of cards falling down.

***

THE RED JEWEL

Sequel

to

BLOOD DICE

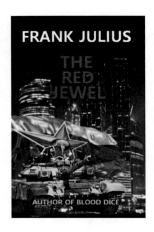

THE RED JEWEL has Eldon Davis expanding his hotel and casino empire into the newly formed Russian Federation. Spurred on by the emerging democracy of the former communist regime, Eldon Davis's Holiday Jewel Hotels tackle his most challenging set of circumstances yet in operating with integrity, while having to deal with The Russian Mafia. Follow Eldon Davis as he takes on his enemies while bringing the world to Moscow through his vision, determination and ingenious solutions to crimes of the century.
A very riveting read of world-class luxury, murder, corruption, sex, violence and service beyond the guest's expectations.
THE RED JEWEL tells the story of the hotel business in the new Russian frontier as you've never known it before.

***